Environmental Education

Environmental Education

Dr. Rajeev Saxena

A.P.H. PUBLISHING CORPORATION
4435-36/7, ANSARI ROAD, DARYA GANJ
NEW DELHI-110 002

Published by
S.B. Nangia
A.P.H. Publishing Corporation
4435–36/7, Ansari Road, Darya Ganj,
New Delhi-110002
Phone: 011–23274050
e-mail: aphbooks@gmail.com

2027

Typeset by
Ideal Publishing Solutions
C-90, J.D. Cambridge School,
West Vinod Nagar, Delhi-110092

Printed at
DIVINE DIGITAL PRINTERS
Ansari road Daryaganj Delhi-110002

PREFACE

Green political theory is a new advent in realms of political theory. It has fascinating as well as tempting avenues to work upon. It is a great mixture of traditional political concepts with the cause of survival of life and plant itself. Infact it is a multi dimentional subject which attracts not only social scientists but also natural scientists, bio-scientists and laymen.

In this parlance, I have tried to sketch the new aspects of green political theory for benefit of professionals as well as students.

I would like to place on record my sincere thanks and obligations to Mr. S.B. Nangia of M/s. APH Publishing Corporation New Delhi for publishing this book in its present format.

I am also thankful to my mother Mrs. Sarla Saxena for her unfailing encouragement and mental support.

Last but not least may lord Mahakal of Ujjain bless this work for cause of ailing humanity

Rajeev Saxena
"Shivratri"
12.02.2010
Utsav, A-37, Indrapuri, Sethingu, Ujjain.

DEDICATED

This work is dedicated to Lord Mahakaleshwara of Ujjain, and Aayush my son, and Eeshita and Arya my nieces

Author

CONTENTS

CHAPTER-1

Basic Concerns of Green Political Theory

"What is Green Political theory" is debated in main stream of political theory for its own existence Traditional political theory considers its field of study in normative issues where as modern political theory underlines its scope in empirical subjects. The normative aspect of political theory searches the values and locates the target for an organised society where as modern political theory reveals the operational truth and empirical happenings behind the ideals. So, ultimately they are the two sides of same coin.

Now Green Political theory is what. It certainly, is extension of political theory as a whole for cause of life and sustenance of this planet and it will be right to say that if life is there beyond earth, the Green Political theory also talks about the same. Andrew Vincent1 opines that one of the deep attraction of Green Political theory is survival of natural. Eco system of the planet is for cause of biological continuance of species and for human's perspective it can be said that where meaningful life is possible.

The Green Political theory goes one step ahead of political theory when it articulates not only the individual and society but also the natural ecosystem, and the survival of not only human species but of other species also.

Green Political Theory has prefix "Green" and the meaning of the same world is simply not vegetation or forest but the whole nature comprising of wild and human and non-human world. The green covers also the socio-economic and political patterns of production, consumption and sustenance of biotic and abiotic life along with the values annexed with this whole gamut. The essense of green this is Nature, que green is the key theme. The indusliing issue of this theory articulates the conditions of relationship between nature and political theory. The political action of human are the action of human are viewed from the angle of holding life vis-a-vis the nature the green is manifested. In fact it is an ontological expression.

The concept of nature has been varidly and widely focused in Green Political Theory. The nature is not only a wild or a complete whole of species. but also it conveys the entropy of whole existence. It comprises of the human centred actions and goals, it also comprises of the action and favour for whole ecology. It is monistic identity having dualistic characteristic of man and wild. The nature is thus an organ one organ which consists of different claims, appeals and reliefs. it also comprises of politics on nature.

"Environmental Politics in about how humanity organises itself to relate to the nature that sustains it. Thus, it encompasses matters of how people deal with the planet and its life and how they relate to each other through the medium of environment. It impinges on other areas of political concern such as those related to poverty, social justice, education, race, economy, international relations, and human rights in as much as what happens in these areas affects our environment (and visa versa)."[3] The traditional concepts of political theory comes automatically under sway of Green Political theory because the inter-human relations which move on these concepts have additional exposure to eco-system and non-human whole.

In Green Political Theory the ecology and environment are studied separately. The former is related with nature having its intrinsic value and man a constituent of eco-net, but the later is related with the managing and correcting environment with human entrepreneurship.[4] This whole view covers realms of environment policy, ethics, and economics as integral part of coverage[5] in political dimension and its consequences.

The concerns of green political theory have been since dawn of origin of society and even beyond that. It is the nature which has nurtured the human species along with others. The Rigvedic Rhymn sings earth as sacred mother and human as its son the very first text of human civilization has this evidence, and this evidence got its attraction internationally in first national people of colour environmental scholarship. Summit 1991, which affirmed the sacredness of mother earth, ecological unity and interdependence of all species and the right to be free from ecological destruction, mutual respect, justice, balanced use of natural resources and clean environment for cause of life.[6] In fact it was call for environmental justice which is projected through these calls is one step ahead of that idea of justice which does not take into second the vote of nature (ecological surroundings)

The green concern of green political theory, thus is not environmental justice but also its sustenance and transformation for future generations. The sense of responsibility thus lies in preserving the rights of future generation. This is the great beauty of Green Political theory but at the same time the operative part of this axion lies that either social justice is a precondition for environmental sustainability or social justice produces environmental sustainability."[7] Sustenance can be inherited only if every natural entity is allowed to enjoy its fullness of its own from life and all life forms are mutually dependent on themselves and non-life forms.[8] And in this light it comes on board that the idea of social justice is instrumental to environmental justice subject to condition that eco-centric values are cherished. The idea of liberty, equality and rights are also examined in Green Political theory but subjection is that econolism remain these. Justice to environment is another epitome of Green Political theory. The rights of vegetation, animal natural resources and even of this planet itself are also the high concern of this issue.

Green Political Theory also contemplates the idea of ecological democracy. Which ensures meeting socio-economic justice to all humans, participation of every walk of life for cause of sustenance of their own life and sustainability and environmental protection of the eco-democracy so designed pawns the path of eco-citizenship for whole of the global in time with" think globally and act locally.

Green political theory has also raised the question of eco-socialism versus eco-capitalism. The co-systemic integrity in the slogan of eco-socialism. The Marxist economics deals with eco-capitalism when it comes forward. Environmental risk stems directly, according the position originally sketched out by Marx and Engels, from the size of capitalism and private property .... As Marx put it . capitalism production turns towards the land only after its influence has exhausted it and after it has devasted its natural question. [11]

Green Political Theory has also concern with market liberalism, globlisation, post-modernism, the rift between north and south bloc, the green movements, environmental policies. [12] And it is not the end of story. The new world order of green political thinkers as envisaged by Arnae Naess, Andrew Dowson, Andrew Vincent, Robyn Eckersley, Robert Guodin etc. is also the concern of the school. The new eco-world order as has been thought of by ecocentricians find its full bloom in Mahtma Gandhi's philosophy. This all is the concern of Green Political Theory.

And last but not least it is advisable to say that every things on this planet and even beyond that it is there for cause of life and its sustenance and continuance is the concern of Green Political theory without fail and beyond any doubt. Every concept which can be determined with ecology and environment is also concern of the debate of green political theory, consequentially.

REFERENCES

1. Andrew Vincent in Richard Bellamy, Manchester University Press, New York, 2003, p. 182

2. Ibid p. 182

3. John Dryzeck & David Schlosberg, Debeting the earth, Oxford 2005, p. 1.

4. Andrew Dowson, Green Political thought, Routledge, London, 2000, pp. 2-35.

5. John S Dryzeck & David Schlosberg op cit. p. 2.

6. Ibid pp. 429-430.

7. Julian Agyeman and others (ed), just sustainabilities, Earth scan, 2003, p. 84.

8. Low & Gleeson, justice, society and nature, Roulledge, London, 1998, p. 155.

9. Julian Agyeman (ed) op cit, p. 93.

10. Daniel R Faber and Deborah Mc Carthy, Neoliberalism, Globalisation and the struggle for ecological democracy in Julian Agyeman and others (ed) op. cit 57.

11. Peter Dickens, Society and Nature, Polity Cambridge, 2004, p. 80.

12. John S Dryzeck, David Schlosberg, Debeting the earth, Oxford 2003.

CHAPTER-2

Green Political Theory A Critique

Green Political theory is an extension of traditional political theory to realms of our ecosphere. This articulation seems to be a very strange phenomens but it is there and it has gained momentum across the whole planet.

The traditional concepts of rights, liberty, equality, common good, a good life, and justice get agglomerated. On a common platform of survival of the planet it also addresses the conditions for our biological continuance as a species, From our own species' perspective green theory could thus be said to be articulating the conditions where by for the meaningful human life is possible".[1] The Green Political theory, thus addresses the very key issue of survival the that element in which the whole concepts of political theory works. This new avenue does not only address the human species and its existence but also the existence of nature in its originality. This is the specific dimension of political theory which has abridged man and non-man on this planet and there exists a specific web of relationship. "Green Political theory conceives of itself as "Green" Environmental or ecological because of its key focus on nature. Nature seems as a crucial entity in its own right of which we are just a very minor part"[2].

Green Political theory takes into account the two players one is human and other is nature or more precisely the non-human every thing around, and how the player one plays its game for its own cause and what are the appeals, obligations and liabilities to the other player are also accounted for. The socio-politico-economical decisions and forces which affect the whole planet is also a big concern of green political theory. This theory also underlines the existence of signature of "nature" in shape of ecology and environment and a complete eco-philosophy. The use of word "green" is symbolic where in specific it concerns for the green vegetative world in general the every thing, event, object and action which happens to be in nature (nature comprises of human and non-human world both).

The green political theory also takes into the account the eccentricities, be it ecology a broader term and notation or environment but the survival of existence is an ontological expression in the same[3]. The existence knows no divide. But anthropocentric issues and eco-centric concern both play in the same cradle. Though the whole earth can be assumed as one organism[4] yet the aspects and concerns have different angle to view the truth. Anthropocentric green political concern weighs every affair on the anvil of utilities and need fulfillment of human, but the ecocentric concerns vetts the whole nature as overall entity and human being the member of the same. Anthropocentric arguments stress that human beings are the sole criterion of value[5] but the eco-centric favour goes to nature as supreme value which is inherent in it. Anthropocentric views may go with traditional political theory where the web of social engineering is studied but in eccentrue aspect the cohesion of man with non-man living and non-living subjects.

The green political theory also talks about the Anthrocentrism: that is suppression and oppression of nature. Out of the greed and materialistic need fulfillment. The issues of conspicuous consumption have arrived with the mere truth of exhaustion or depleting natural life sustaining teasure. The life has come in big danger and it is approaching toward extenstion. This all has taken place velemently in capitalistic mode. The socialistic patterns are also taking the capitalistic mode and exploiting the natural resources. Over population or increasing demographic pressure is causing the exploitation of nature. Besides it all the five elements basic to the study reported have population, food production, industrialization, pollution and consumption of non-renewable natural resources are increasing.[6]. These all elements have caused exploitation as well as tyranny over the wild, and socio-economic forces and political motives have been instrumental in this whole story. Industrial output is growing at seven percent per year and population at two percent. This equation has put the day-by-day widening non synergy between rich and poor developed and developing, north and south, have and have not[5] right and left, etc. The process of economic growth as it is recurring today is inexorably widening the absolute gap between these two hemispheres and this is ultimately causing androcentrism. All these factors vis-à-vis ethics of development is also a concern of green political theory.

In the same tune when the wild is at the verge of collapse and nature is at the edge of destruction, the nature of "nature" is also a big concern of green political theory. "Nature is the sum of what is not the result of human action." Inimically this idea has been attractive for ecocentric theories."[8] Deep ecocentrists understand human as villain. For them nature should be untouched by man. Man is simply a knot in the big web of nature. The characteristic of nature is dualistic; it envisages nature at one end and man at the other. In light of eccentric theories, nature is a bio-spherical net and human is simply a knot in it. The other broad sense regarding nature is "sum total". The sum total includes humans and all their actions. If this views is taken then the greedful and lusterous actions of human and thus environmental degradation will have to be accepted, and further more in this light the relevance of "greenism" will lose its ground. The dualistic concept of nature and the monistic concept of nature as narrated supra form the basic ground work in green political theory. Though out of this dualism some says that the "concept of nature is deceptive" [9] Yet this dualism confirms the further travel of green political theory. "The upshot of both these views is that there are certain forms of social and political arrangements which are either harmonious with or functional for the natural environment. Consequently it is possible to identify a natural sense of nationality, democracy, citizenship or justice. However, given that nature is contested. What effect does this have on green argument".[10] This view incorporates ecocentrism, anthropocentrism and a soft and subtle androcentrism.

The philosophical stand of green political theory is a sum balance of idealism and empiricism. The green theory ethics dealing with socio-politico-economic ecologism comprises of the deep ecologism but with a minimum self sustaining need fulfillment requirement with simplicity and austerity. The eco-centric ethics can not be accepted by anthropocentric and vise-versa but a bridge between these two hemispheres can be explored in shallows eco-centrism – It accepts our. favouritism for our own conduct; adopt and modify our activities .... The ethical standing of nature is itself natural. [11]

The nature thus has its own intrinsic value. The nature can also be a means or an end in green political theory. 'Nature' untouched by man or touched nature having level of 'culture', have got special appeal in green issues. When we talk about naturalism, it means either we follow and advocate community for life sustenance or natural ways of living without disrupting not only green but every thing non-green around. It this form, ethics is therefore, viewed in form of communitarian naturalism.[12] But what is the limits of communitarian ethics, it is blurred. The global ethics also stands before us and it involves various communitarian patterns or it is an universal ethics of not only sustaining green values but also the natural values. The nature consists of not only the green but everything which is not-green and when the green political theory is examined in the context of word nature, the horizons become incessantly infinite.

Thus the crucial question is who or what defines nature – whether it is culture the man made effect or the evolutionary nature which accompanies the man with its development and action. The whole issue baffles. Moreover, the green political theory which emerges from the necessity of sustenance of life discusses the issues of man and non-man both at the same time. Green political theory thus not only analysis the philosophical stand of ecologism or environmentalism along with its off shoots but also weighs and marks the naturalism. Some[13] may perceive the teetring on incoherence of Green political theory just because of baffled question of nature of nature, but ultimately the question of green political theory philosophically stands for a common good life inheriting to coming generations also without disturbing the natural web of whole existence.

A theory of sustainable life that is green political theory is beyond the realm of mere politics or socio-economic forces and touching the essence of value of right livelihood.

REFERENCES

1. Richard Bellamy and Andrew Mason, (ed) Political concepts, Manchester University Press, Manchester, New York, 2003, p. 182.

2. Ibid p. 182.

3. Warwick fux quoted in R. Syluan, A critique of deep ecology, Ratical Philosophy 40-41 (1984-85), p. 103.

4. Lepold quoted in D Scherer and T. Attig (ed), Ethics and Environment (New York) Prentice Hall, 1983, p.7.

5. Richard Bellamy and Andrew Mason, op cit, p. 182.

6. Dunella & Meadows, and others the limits to growth (universe books, New York, 1972, pp. 25-44).

7. Op. Cit 25-44.

8. Richard Billampy, op. cit, p. 187.

9. Ibid p 188.

10. Ibid p. 189.

11. Ibid p. 109.

12. Ibid p. 190.

13. Ibid p. 193.

CHAPTER-3

THE "GREEN" IN GREEN POLITICAL THEORY

The word "Green" connotes multifarious and subtle meanings in green political theory. It is not the colour of negetation and it certainly, is not the philosophy of afforestation or clearing environ mental problems. The relevance of word green is not due to text that it has underlined the contemporary concerns of green management but it has again revised a whole alternative socio-economic and political philosophy for sustenance of life. It has been rightly said that the so called greening of the society is the attempt to incorporate the cost of living into our economics the totality of existence into our ethics, and the sharing of the living word into our politics[1]. And one step ahead to this version the meaning of green is not only comprising of the "human", it also consists of the "Nature" as a whole organism with various explanation regarding nature of nature.

Greenism is not the particular "ideology" or any 'ism' as happens to be an traditional political philosophy but it is the whole way of living a life and accentuating the "good" of life being in tune with pursuing the basics of nature.

Green political theory is a peculiar environmental theory which consists of "green" – as a subtle relationship between man and non man. As Robyn Eckersley comments that these fundamental issues have rarely been the given of day in contemporary political theory. Inter human relations take absolute priority in main stream political theory. The state, sovereignty, justice, equality, rights, and freedom are seen to be focused unremittingly on humanity, as indifferent to nature[2]. Thus "green" is the sense of comprehension of the inter-relation between human and beyond human world and the theories which define the intra-human relations are also seen with a new vision of holistic existence.

Thus "green" can be sensed broadly in three shapes Deep green, shallow green, and dry green.[3] Each green is not a peculiar set of person but it is a specific pattern of societal values.

Deep greenness is ecocentric structure, It favours the self reliance, and self-sustaining socio-political communities. It cares for the nature as a whole organism and men as an entity. In name of its radical orientation it does not support the industrialization and the economics of consumption. It certainly not favours the capitalistic pattern of production and consumption. This greenness also cares for fair distribution of fruits among generations. It emphasis the ecological ethics of maintaining the wild in its originality. It can also go against any technical help which disturbs the nature.

Shallow greenness is environmentalism but the ecocentric values are there. It is supportive of technical support and management to environmental problems. In heart of this philosophy the fall respect to nature including wild is there and there is eccentricity to save it from degradation but the human entrepreneurship is respected here.

Dry greenness is connected with scientism and technocratism. It involves the environmentalism with aid of scientific efforts. It believes in complete management of society with human efforts. For this school the nature and man are two actors.

Deep green, thus acts for a total way of ecological life and dry green is totally in favour of managing ecological and controlling the nature with technical support shallow green is a way between.

Deep greens support the green rights as well as animal rights. The value of nature is intrinsic in this school. It supports the global environmental sustaining programme. Shallow greens support the rights of females and they regard the adjustment with the environment and needs of the human. Dry green favours the consumerism and thus it is a total management of environment.

Deep greens believe in self-regulating local communities, shallow greens believe in inter national federation which cares for environmental problems, but dry greens favour the central national power to decide the environmental issues in new forms.

Deep greens support peace, security and fair distribution of socio-ecological fruits not only to prevailings generation but also to coming generations. It's beliefs are spiritualism.

Shallow greens support all that is supported by deep greens but with the more concentrated political power i.e. state involvement. International cooperation and movements are there for cause of saving the environment to become hazardous and inimical to biotic life.

Dry greens support the shallow greens but with self-controlling and sovereign nations at their on own stand. International adjustment through meetings, conferences and protocalls have a little say.

Thus it will be seen that journey from deep greens to dry greens is a journey of concentrated green wild to diluted wild and concentrated naturalism to diluted naturalism.[4]

In short it is a journey from ecocentrism to technocentrism, earth centred or nurturing mode to human centred or manipulative mode.[5]

The ecocentric or deep green political theory focuses the nature and absolute priority of nature, however the traditional political concepts are supplemented by not only claims of human world but also coloured with the supremacy of existence – the nature as a whole.

Ecocentrism to anthropocentrism is the way of pendulum motion in deep green to dry greens. In this way deep greens have been accused of being eco-fasists[6], who sacrifice the human and its liberty on the altar of ecological balance – the nature in its full bloom.

The "green" in political theory spreads the whole spectrum of bio-socio-natural web of relationships. It is not only indicative to keeping nature unpolluted but also the traditional political concepts of state, sovereignty, rights, liberty, equality and justice intermixing with the existence of nature for cause of sustenance of life. The gradual increasing environmental hazards in shape of ozone layer depletion, carbon emission, acid rain, deferestatioon, climate change, increasing sea level, melting glaciers, effects of pesticides, pollution of air, water and land, toxication of edibles and vegetation, extinction of different species, exploitation of natural resources up to limit of non-rechanging and complete inbalance of biodiversity etc. have redefined the traditional political concepts for cause saving the life. Liberalism can not be liberated democratically so that the basic life of this planet comes to an end, and for saving the life the fascism of green in permissible, without any doubt.

"Green" has also the meaning of eco-political ethics, the very ethics which fosters the life by entacting the natural resources, and in this way it also redefines our mode of production and consumption. In fact "green" is a holistic determinative which reshapes our political theory in its totality that is green political theory.

It is also interesting to note that green pohliseal theory has got the previlege of adding prefix of "eco" before any traditional political concept and when it is done the "green" of green political theory comes into the existence.

In fact at last all the five important aspects of nature[7] forest, land, water, biodiversity and air, come consequently in realm of "green" in which the effect is to be studied.

REFERENCES

1. Agarwal, SK, Green Management, APH Publishing, New Delhi, 2005, p.1.

2. Richard Bellamy and Andrew Dawson, Political concepts, Manchester University, Press, 2003, p. 186.

3. O, Riordan, The new Environmentalism and sustainable development, The science & total environmentalism, Pion London, 1991, 108; 5-15.

4. Op. cit, 108 pp 5-15.

5. O. Riordan, Ibid pp 5-15.

6. Tony Lynch and David Wells, 'Non anthropocentrism? A Killing objection, Environmental Values 7:2, 1998, 158-63.

7. Samir Bagchi, towards an alternative paradigm of development, in Biswajeet Mukherjee (ed) Environemtal awakening, Shikha Books, Kolkata, 2005.

CHAPTER-4

The Nature in Green Political Theory

Political Theory deals with the basic entities of political life. The philosophical foundation of polity political instruction, and political theory actors come in the purview of political theory political theory is analysed by the angles of metaphysics, epistemology and axiology. How the state and its citizen acts mutually and on what civil issues, are studied in political theory. In this context green political theory focusses itself on the very survival of whole natural ecosystem of the planet.[1] Green political concerns, thus have three angles of concern! Traditional concepts of political philosophy[2]. Nature and its existence, and[3] inter and intra action between the above two.

The nature which has got the central position in green political philosophy has a contingent concept in history of political philosophy. In Greek tradition, "nature" is an intelligent ordering of the God and it is purposeful[2]. This idea was influential in mediavel political though. In modern times the evolutionary idea of nature and to it man is an integrate part who has adopted itself according to natural evolution is the central issue of thought. This idea further elaborated the stream of thought that nature has its own intrinsic value for evolution and for the cause of survivalism. Conservation of nature is an inherent charactistics in the way which underlines permanent support for the conservation will depend on the success of an appeal to a deeper feeling for nature as something valued for its intrinsic interest and inspirational significance. People need to be pursuaded to respect nature for its own sake[3].

But we know that nature is not devoid of man and its action. The human world and non human world both encapsulate the nature. Nature has intrinsic as well as instrumental value. The nature is analysed in green political concern form two angles.

1. Ecocentric angle

2. Anthropocentric angle

Ecologists provide a rationally different version of nature and the place of the human being within it, one that is 'eco-centric' or nature centred rather than anthropocentric[4] and anthropocentric doctrines are human centred. The ecocentric version puts the nature as an end." The radical ecocentric perspective of writers such as Arne Naess, Bill Devall, Warvick Fox, and Robyn Eckersely reads" that there is no firm ontological divide in the field of existence.[5] Here the monistric conception of nature in adopted.[6] The nature has its own value as an ultimate end and the human is an integral part of this whole organism.

Another Angle for viewing the nature is anthropocentric. Anthropocentric argument stresses that human beings are the sole criterion of value.[7] The nature can be managed or arranged as per human needs, derive and attitudes.

These two angles put nature in center in the first one the nature and its virginity as monistic organism can not be disturbed but in other the nature can be arranged and worked upon as per human values.

The nature when is seen from the anthropocentric perspective is manageable as per Andrew Dobsen "Environment which is manifestation of nature can be managed without fundamental changes in present values or pattern of consumption[8]. The human is supreme in this way then nature. Here the non human all the environment can be manage, but in eco-centric perspective the natural world is as an inter locking system of interdependent objects (both sentient and non-sentient) generates a sense of equality, in that each item is held to be necessary for viability of every other item. In this view, no part of the natural wall is independent and therefore, no part can claim to superiority. Without the humble bacteria that clean our gut wall, for example, human beings would be permanently ill.

It is also to been that deep eco-centric values do exist with intervention of human values to nature. If human is integral part of nature than its actions (be they are coming out of needs, derives or greed) are also natural. "The sum total includes humans and all their action... to follow out its logic more rigorously would include all human activity in industry, economics, culture and politics within the ambit of nature. This is turn, raises a further issue, namely that environmental degradation, as a result of human actions, could also be considend natural. If humans are an evolutionary species, then all human activities are natural; even if some are extremely rusky for species survival."

But if this freedom of naturality is taken then the stress of earth like rising temperature, melting ice, rising sea level, destructive storms, dry rivers, falling water level, water security." Collapsing of fisheries, shrinking forests, deterioration of rangeland, soil erosion, disappearance of species, synergies and surprises[11]. Emission of CO_2, depletion of ozone-expanding deserts, over pumping of aquifers, over flowing and overgrazing, environmental changes bringing food insecurity etc are unavoidable to be reached at.

It is natural and if it is natural. Our extinction from survival platform takes place, this analysis frames the question again what is nature and what is nature of nature.

If all these stresses which are there are seen from anthropocentric angle of nature, the same situation is not a problem till the maximum limit of sustenance is not crossed.

Whether nature has different meaning in different perspectives, but it cannot be the truth because a quality life continuation is essentially a natural law.

There is a third angle to sea the nature which is pliant anthropocentrism "Pliant anthropocentrism stresses co-dependency with nature although still fillered through human interest"[13]. When analyzing the nature in green political theory other basic factor that is, increase in population, industrialization, pollution and consumption of non-renewable natural resources are additionally adding up the nature[14]. Green political theory is questioned by these additional factors considerably.

The liberal democracy, market liberalism globalisation and hyper industrialization have tilted pliant anthropocentrism. Towards anthropocentrism. The situation is that natural resources are running out. The population is ever growing. Leaving less to eat, species are becoming extinct in vast numbers : forests are disappearing and fish stock are collapsing, The planet's air and water are becoming even more polluted[15].

So it is clear that nature is under continued stress be it eco-centric assuming condition that man is natural and is actions are natural), anthropocentric or even pliant anthropocentric. "The wild naturalness is not possible in all these because if there is man, he will do some thing according to his value structure. Against this picture the eco-logism which is an green political ideology" starts not form a conception of humanity or human needs, but from a vision of nature as a network of precious but fragile relationship between living species including the human species and the natural environment. Humankind no longer occupies center stage, but is regarded as an inseperable part of nature. Human beings are therefore required to practice humility; modernations, and gentleness and to abandon the misguided dreams that science and technology can solve all their problems."[16] Back to nature theme is manifested here. The basic question is that whether it is possible, when developmental economics is on its boom. Though moral questions on sustainability of developmental has been raised but the nature in its whole entity has continuously been squeezed. The managerial aspect of environmentalism as quoted by Andrew Dobson has been envisaged as amicable solution to approach the virginity of nature.

The present stage of nature also concerns with the coming generation. The nature which will be interacting with forthcoming generation should dispose off the distributive justice and this may take place so the concept of sustainable nature is now taking place. Sustainable development, which cares for weaker section in process of development takes into account the claims of forthcoming generations, and it should be substituted by the sustainable philosophy of development read with the survivalism, The millon dolar question is that if green theory does articulate the conditions of ecological and biological survival and flourishing, then politics must be umbricated, in the sense that how humans act politically has a crucial impact on nature and, thus indirectly upon our survival as a species[17]. In this partance, the characteristic of nature in green political theory comes out that non human world is quite rich and vital to present and forthcoming generations for the cause of survivalism, the intrinsic value of nature is achieved, At this juncture eco-centric or anthropocentric angles leave their ground. This also devotes that, "survivalism" is the key factor for defining the "nature" of nature in green political theory.

Nature in green political theory is also deceptive when we talk about ecological justice. Ecojustice is distribution of natural resources and with mutual respect between human and non human words therefore environmental theory has mostly tended to assume that ecoharms are generalisable, affecting all people, within an abstract national community more or less equally and ecological rehonality should be approached therefore through a politics of the "Common good"[18]. The common good is another key factor for defining the nature of nature in green political concerns. This common good is not anthropocentric but it, certainly is eco-centric where not only humans but rivers, field, mountain and biotic world may also feel good.

Nature in green political theory is also linked with flourishment of culture and economics. Our traditional laws lead us to understand that economic development cannont subsist on a deteriorating resource base. The environment cannot be maintained and protected when "growth" does not account for the cost of the environmental and culture destruction"[19]. If the cultural destruction is appearing in nature, that nature is not the key of green political theory. This issue was analysed by social ecologist Murray Bookchin who involves subtle blending between nature as distinct and integral to us. As a biological evolution human species has come to this stage. This is one aspect of nature. Another evolution is social and cultural evolution. This is another aspect of nature. Evolutionary process takes both the aspect and, "As we evolve, we see ourselves as nature rendered self conscious and intelligent: In social ecology we co-operate with the implicit teleology of nature[20]. and culture can thus be seen a collective capacity of humans to adopt to the particular and contingent conditions of their collective existence including most importantly the environments with which they interact and upon"[21]. Thus in green political theory that is not natural characteristic of nature which cannot support the culture. Deterioration of culture through environmental degradation is nothing but the nature's appeal to save it.

Green political theory keeps the nature in the center. The survivalism, common good and cultural maintenance is its utmost. and subtle sense conveys the teleological doctrine of nature. One more thing should be remembered that human and non human world both have their rights on nature in this sense.

In the same tune it is also to be noted that the perception of nature has been thought of in following manner[22].

1. Every thing in nature is connected

2. Nature is kind and fulfills the requirement, provided we do not disturb the natural ecosystem.

3. Nature is fragile and if we change nature, it proves disastrous.

4. Nature is durable. It means science and technology and human efforts can reshape the nature. The damage

control to nature can be done through auxilliary methods.

5. Nature is capricious. It means nature is not certain. At time it may prove benign and on the other moment it may prove perverse.

The above reflections again divide the nature in eco centric, anthropocentric hemispheres. Whether the nature is vergin and must remain the same or it be touched and used is the question of million dollar. If it is to be vergin, humans should be part of it and they are knots in web of nature. In this sense disastrous activities of humans also prove natural, but on the maxim of "survival", such activities are not welcome. If it is to be touched, it be done under control limits.

The point is that "nature" is a mystical term to green political theory but the assertion of "nature" is most attractive at the same time.

REFERENCES

1. Andrew Vincent, Green Political Theory in Richard Bellamy and Andrew Mason (ed), Political concepts, Manchester, university Press, New York, 2003, p. 182.

2. Andrew Vincent op. cit. p. 188.

3. W.M. Adams, Future Nature, A vision for conservation, Earth an Publication, London, 2003.

4. Andrew Heywood, Political Ideologies, palgrave Maemillamy, 2003, p-271.

5. Warwick Fox, quoted in R Syl van 'A critique of deep ecology, Radical Philosophy 40-41, (1984-85) p. 103.

6. Andrew Vincent op. cit p. 183.

7. Op. cit. p. 184.

8. Andrew Dobson, Environmentalism and ecologism in Alan Finlaysun, contemporary Political thought, Edinburg University Press 2003, p. 364.

9. Op. cit. p. 380.

10. Andrew Vincent, Ibid p. 188.

11. Lester R. Brown, Eco-Economy, building an economy for the earth, earth scan, pub. Ltd, London, 2001.

12. Lestey R. Brown A planet under stress, in John S. Drybeck David Schlosberg (ed), debating the earth, Oxford University Press, 2005.

13. Andrew Vincent, Op. cit p. 184.

14. Donella H. Meadows, and others, the nature of exponential growth, in john S Dryzeck & David Schlosberg, op. cit, p. 11.

15. Djorn Lomburg, The Truth about he environment, in John S. Dryzeck and David Schlosberg (ed) op. cit, p. 74.

16. Andrew Heywood, op. cit, pp. 268-269.

17. Andrew Vincent, op. cit, p. 182.

18. Val Plumwood, Inequality, eco-justice and ecological rationality (in) John S. Dryzeck and David Schlosberg (ed.) op. cit p. 619.

19. Walt Breseile et. al, Seventh Generation Amendment Anishineabe Ninji flyer, Bay field, WI, March 1996.

20. Andrew Vincent, op. cit, p. 190.

21. John Berry, Rethinking Green Politics, Sage, London, 1999, p. 50.

22. Gerald D Marten, Human Ecology, Earthscan Publication, London, 2001, pp. 122-125.

CHAPTER-5

The Earth in Green Political Theory

The green political theory envisages the cause of life and its sustenance for present generation and the natural continuance of this planet for a better life of forthcoming generations. We have analysed the "nature" and "green" previously. The green political theory can't move further by these two elements. But the perspective of "earth" is equally important in green political theory "Nature" and "green" have "earth" in them it is univocally accepted. It is the earth which sustains the life and other 'things', 'world' and "non-human world".

In the magnacanta of our environment.[1] at stock home declaration, the natural resources of earth including air, water, land flora, and fauna etc., have been ordered for keeping safe for benefit of present and future generation, and at the same time the vital resources of land be maintained. Despite of open political-economy and liberalisation and even capitalistic modality of production and consumption, the earth and its limits have been kept in focus. These dictates underline the political action on earth. Political agendas, issues and system, thus have been warned through these principles to not to scratch earth for end of non repletion of its treasure.

In the same way principles of environmental justice[2] adopted at Washington in 1991, the earth has been affirmed the sacredness of mother earth and to keep safe the vital resources of earth for present and future generation. In this adoption the earth has been given status of mother and sacredness has been assigned to it. It is the repetition of the hymn of Prathvisukta of Atharvaveda "that the human being is a part of nature, may the very child of earth[3] . The ethical action on body of earth through & super fine rather divine feeling is the core idea. Which will guide control and curb the political and economical action.

In the same way the conferences of Rio 92, Johannesburg 2002, and earth summit 2003 propounded the theory that earth is a precious entity and no more lust full action on its remaining virginity is required further more. Even the liberal economic activities in the contemporary free political societies have been restricted, as this plunders the earth.

The eco-centric view of green political theory has sough to cultivate human solidarity with non-human nature by inviting us to expand our identification with the world around us, to consider the rest of nature part as of our expanded self[4]. In this way earth is nothing but an expanded self of humans themselves and as a natural law, no one can kill or hurt himself or herself as the earth can't be hurt in any way be it politically or economically.

The sacred ness of earth thus limits the rights, equalities and liberties of human and nations in the name of democracy and sovereignty respectively. The check on the political concept of sovereignty is implied upon by the "earth" that is it because it happens to be the sustainer of life.

The political character of earth is coincided with "Gaia" hypothesis (Gaia – from the Greek Goddes – 'Mother Earth') which reads in a balanced way the humocostasion earth as planet. The balance in bio-sphere and interdependence of natural objects[5] is there. "Gaia" hypothesis is an working ethics to earth. This is a socio-political and economic ethics to deal with the earth. In fact "Earth" comprises of "nature" up to extent of natural homoeostasis.

The earth has also got its full say when we talk about the eco-citizenship that goes as the obligation of ecological citizens knows that today's act will have implications for tomorrow's peoples, and will argue that the claim of future generations are to be kept safe[6].

The "earth" in green political theory has normative values and ontological expression. It has got its own existence and so the claims and rights. The rights of earth are to keep it clean, green safe and divine with full of vitality so that life máy continue on it for ever. The earth has its intrinsic value in itself. It does not get valued by man. Man is the part of this sacred mother.

The political theory when talks in reference to earth as a being it puts on the rob of political theory of earth. The use and misuse. Of earth through is a political action a big concern of green political theory, and 'earth' has got the center place in this.

The philosophical basis of "earth" in green political theory is not like an insert substance, that can be used in need of human, rather it has got central position of game when "earth" puts on the great roll of "greater earth" – that comprises of not only subsistence of life but also the nature as a whole entity. In concrete terms "earth" is an element of environment but in abstract term "earth" is upholder of truth of vivid life and its causes and in this way the green political theory is ultimately the theory of life sustenance; earth nurtures, nourishes and flourishes and transmits the life and the life sustenance.

Sustainable earth theory is that which supports bio diversity and fulfills biological goals along with the betterment of life and this process sustains the present generation as well as present with future generations.

REFERENCES

1. Stockhom declaration, 1972 in Biswajeet Mukherjee (ed.) Environmental awakening Sikha Books-Kolkata-50, 2005, Principle 2 & 3, p.1.

2. Principles of environmental justice, adopted at the first, national people of colour environmental leadership summit, Washington DC Act 1991.

3. Ranjan Chakrabarti, does Environmental history matters, renders service, Kolkata, 2006, p.90.

4. Arnac Naess "The Shallow and the deep long-range ecology movement " A summary, inquity 16(1973), pp. 95-1000.

5. Prabuddha Nath Rag, Ecology, Economics and sustainable Development (in) Biswajeet Mukherjee (ed), environmental awakening, Sikha Books, Kolkata, 2005, pp. 96-99.

6. Andrew Dowson, Ecological citizenship, Oxford University Press, 2003, pp. 83-140.

CHAPTER-6

Ecologism – The Core Values

"Man" and "Society" and the interacting dynamic social sources have been the vital concern of socio-political philosophy up to mid of twentieth century. In post mid period of twentieth century another factor of surrounding environment itself appeared on the canvass of thought. This factor defined the value of survival of the whole natural system with respect to "man" and "society". The new arrival underlined the basic of environmental importance and existence of human vis-à-vis the balance of nature and thus the ecology. The virtue of ecologism is that in stressing the central importance of ecology, it highlights an approach to political understanding that is quantitatively different from the conventional ones. In calling for radical socio-political change and a fundamental retheking & the relationship between human beings and the natural world, ecology has developed into and ideology in its own right[1].

The concerns of ecology, though, owe much to oriental philosophies of Hinduism, Buddhism and Taoism, where mother earth was the idle of reverense and sustenance[2]. The Vedic thought echoes the earth as mother and the human species as if the son of this mother. The process of growth and industrialization and urbanization that was seen in the post mid twentieth century provoked the enquiries of cost of material progress and exploitation of natural resources, and the relationship of "man" with environment. The affluency of material and scientific progress questioned by virtue of its own existence, the value structure of development. Continual debates, thoughts and issues of pollution, conservation, acid rain, green house effect, global climate change eco-political issues and human species existence evolved the green theory in general and concentrated the green political theory in particular.. The deep attractions of green political theory is its claim to be focused on the very survival of the whole natural eco system of the planet. In consequence it also addresses the conditions for our biological continuance as a species[3].

Other political ideologies analysis the socio-politico-economic engineering but the eco-logism concentrates on the core most part of human existence the basic "unit" of planets and the governance of the same. The philosophy of ecologism does not focus human needs only but also takes the nature-the cradle of humanity in its sphere of analysis. Thus there are two basic thoughts one is ecocentric and other anthropocentric. A balanced eco-logism talks about the both. Hard liners are there but ultimately they also resolve on same amicable issues. Arne Nases, Bill Devall, Warwick Fox and Cobin Eckersely[4]. are eco-centric. :The eco-centric value perspective has developed on two lines. The first is intrinsic value theory, which sees nature an end in itself... The second eco-centric perspective bypasses value theory. It argues that what is required is not so much ethics as a psychological change in ecological sensibility. The real issue is therefore psychology and ontology not ethics[5]. Eco-centric drive their strength from nature and a peculiar respectful reverence to the same. The nature whose part is human also, but ultimately the nature itself is the central theme of this school. The value structure for this school emanates from the virginity and piousness of nature. Human desires and need, thus, certainly, are subject to the same virginity. The governance and so the political theme of society should be guided from this angle only. Another anthropocentric thought opens that the human beings are the sole criterian of values. The endeavors of human and needs and then fulfillment is the central theme of anthropocentric thoughts. "Deep anthropocentrism is indifferent to nature and is largely outside the domain of green theory. Plaint anthropocentrism stresses co-dependery with nature although still filtered through human interests[6]".

Hard liners and soft liners both are there, for some nature is supreme and for other sote human need and its fulfillment. Deep ecologists, deep greens and shallow ecologists and light greens, both are there : environmental ethics, thus markes the drift from the personal ethics, rather it operates for whole planet by respecting the vegetable vitals, mineral vital, natural flora and fauna vitals, natural global elements and the human as an integral part of this vast treasure.

The central themes of ecology are the following[7]:

1. Ecology
2. Holism
3. Sustainability
4. Environmental Ethics
5. Self Actualisation

The ecologism is hard ecocentrism. The man is not the master of nature but an integral part to it. Deep ecolgists advocate biocentric equality, animal rights, nature, value in nature, antigrowth anti ecological consciousness whereas shallow ecologism is light anthropocentricism. Shallow ecologists advocate conservation of non-human nature, animal welfare, substantial growth-sustainable growth and personal development. It is like: deep ecologists speak of wilderness. It is usually nature untouched by human action. Humans are envisaged a mere temporary knots in the bio spherical net"[8]. but on the other hand there is crisis of assumption, if the nature is a sum total, a complete whole then the huan species and its action by being of an integral part of the nature are also natural, even exploitation and environmental degradation due to human intervention are justified. But this assumption can not be justified. An order which happens to be in untouched nature can not be broken. The bio-natural balance is essential. From which point the intervention of "humanity" impedes the goals of ecocentrism, is also not clear only the ethical way describes it. The ethical position of "human" in whole nature is that in biocentrism the freedom is there for human to play but decodes how to reach given environmental goals and at the same time the human is not privileged.[9] In this light the rights of plants, vegetables, animals and every other constituent of nature find its strength. It then becomes natural manifesto of nature rights to every creation. It can also be seen that the question of ecocentrism can be weighted against liberalism and socialism. "Liberalism presupposes the human indivudual as its basic unit; socialism presupposes human collectivities; but ecologism presupposes that human are part of nature"[10]. Human is liberal and social but it is eco-liberal and eco-social creature in eyes of ecologits.

The term holism describes the idea that natural world can only be understood in its wholeness not through its individual parts. Capra[11] rightly in this sense draws the picture that alike oriental religious doctrines of Hinduism, Buddhism, and Zen Buddhism. There exists oneness of all the things. The theory of quantum physics teaches the all related in one and thus theory of everything (OE). It points out towards a refined concept of ecologism having "human" as one of its constituent, but ultimately this trend is in favour of deep ecologism. Sustainability is another aspect which is bricked with economic progress vis-à-vis ecohumanistic activites. The non-human nature of planet is like a stuff of space ship earth. It has limited and exhaustable resources. Reckless exploitation and violent behaviour has degraded and spoiled their sanctity. Living in a spaceship requires an understanding of the ecological processes that sustain life. Most, importantly, human being must recognize that spaceship earth is a closed system"[12]. The closed system has the entropy character of energy. So in closed system the transformation of energy takes place but the form of energy resources get exhausted if a merciless treatment is done. The bio-spherical egalitarianism is advocated for a complete concept of ecologism.

Environmental ethics and self aclualisation are also important aspects of ecologisms. For a good life a balanced civic society is essential. Here lies the civic consciousness, the ethical values and undoubtedly a political ecology. The "human element" and "non human element" both must be at certan e cological matrix mutually." Political ecology is a historical outgrowth of the central question asked by the social sciences about the relationship between sciences about the relationship between human society. Viewed in its bio-cultural political complexity and a significantly humanized nature[13].

The crisis of faith in the values of ecologism, thus falls in the following argument.

"Environmentalism argues for a managerial approach to environmental problems, secure in the belief that they can be solved without fundamental change in present values or patterns of production and consumption, and ecologism holds that a sutainable and fulfilling existence presupposes radical change in our relationship with the non-human natural world and in our mode of social and political life"[14].

Ecologism underlines the math of traditional concepts of political philosophy with the ecology. Justice, rights, equality, liberalism, socialism, democratic values and action, post-modern values, hyper industrialization etc. all are taken care under umbrella of ecologism. The relationship and the distribution and association of their values between biotic and abiotic world has been defined in ecologism. As mentioned above the environmentalism is a managerial approach which keeps anthropocentrism duly balanced with ecocentrism, but the ecologism keeps human race as an integral part of the whole "the whole" is to be taken care of. It is apparently clear that the value structure, then gets change in their array. Contemporary age is of democratic values, effective participation, gaining enlightened understanding, exercising final control over the agend[15] attract care for survivalism. The common good of Rousseau is concern of these democratic agenda with ecological balance. The questions of values imbibded in conservative, liberal, socialistic, capitalistic or blend of different political economies, ultimately point towards a better life and survival. The ecologism and the holistic values attached with this stream of thought go beyond the realm of traditional political theory. It gives respect each creature and creation, biotic and abiotic world simultaneously. The schism in values between ecocetrism and anthropocentrism can thus be balanced on the axis of just relationship between biotic and abiotic world; ultimately the survivalism of all.

Social ecologist Murray Bookchin asserts, that a subtle blending between nature as distinct and nature as integral to us in essential "nature rendered self conscious and intelligent" is cooperated with the implicit characteristics of teleology of nature[16]. The ecological movements are in fact political movements for a non-violent order in which nature is conserved for conserving the option for survival. They demand only the right to survival, yet with that minimal demand is associated the right to live in a peaceful and just world... Unless the world is restructured ecologically at the level of world views and life styles, peace and justice will continue to be violated, and ultimately the very survival of humanity will be threatened.[17]

It is clear that the ecologism as a school of thought conserves, ultimately, the value of life and the sacred rhym, live and let live with all the essentials of minimal socio politico economical and sustainable development that will not lead to bubble economy.

REFERENCES

1. Andrew Heywood, Political Ideology, An Intraction, palgrave Macmillan, New York, 2003, pp. 216-217.

2. Ibid p.-217

3. Andrew Vincent, Green Political Theory, Richard Bellamy and Andrew Mason (ed), Political concepts, Manchester University Press, New York, 2003, p. 182.

4. Arne Naess, Ecology, community and life style D. Rothenberg (Cambridge University Press, 1990), Bill Devall and G sessions Deep ecology, living as if nature mattered (Salt lake city UT Gibbs

M Smith Inc, 1985) R Eckersley, Environmentalism and Political theory (ULL Press, London, 1992)

5. Andrew Vincent, p.184.

6. Ibid p.184.

7. Andrew Heywood, op. cit, p. 269.

8. Andrew Vincent, op. cit, pp. 187.

9. Stig Warnden Ethics Prices and Biodiversity Sweedish environmental protections Agency, S 106, 48 Stockhom, Vol. 14-N-1, 2001, p. 96.

10.. Trim Hayward, Ecologism and environmentalism in Alan Finlaysen (ed) contemporary Political thought, Edinburgh University Press, 2003, p. 353.

11. F. Capra, The Tao of Physics, Fontana, London, 1975.

12. Andrew Heywood, op. cit p. 277.

13. Greenberg James Band Park Thomas K. Political Ecology, Journal of Political Ecology Vol. 1, 1994 p.1.

14. Andrew Dobson, Green Political though, Routledge, London, 2000, p.2.

15. R. Dahal, Democracy and its critics, New Haven, CT, Yale University Press, 1989, pp. 106-31,.

16. Bookchin, Towards a philosophy of nature, in M Tobias (ed) Deep Ecology (San Diego, CA Avant Books, 1985) p. 228.

17. Vandana Shiva, Development Ecology, and Owmen, in Andrew Finlayson (ed), op. cit, p. 437.

CHAPTER-7

ENVIRONMENTALISM : THE ETHICAL BASIS

Recently up stood stream of great debate of Green idea has stirred the minds of world humans. The ethical foundation of right and wrong, while interpreting the environment, at the same time, added more colours. The degrading environment and threat to human existence vis-à-vis the planet provided an additional momentum to the ethical foundation of environmental issues.

The environmentalism which "argues for a managerial approach to environmental problems, secure in the belief that they can be solved without fundamental change in present values or patterns of production and consumption.[1] But this vies has been critically attacked by ecologists who are of opinion that present value patterns of distribution and consumption is detrimental and unethical to whole existence. The laters pose the bogey of questions before the values and beliefs of the former. Economics decides the ethics, it is a well-established dictum of Marxian politico-economical theory, here in the environmentalism, the factors of production distribution and consumption underlines the inherited sense of mastery over nature – a slave like look at nature, and it, really has taken place "In affluent countries the post-war period was a prospect for everyone, as the fruits of technology were being harvested and nature was exploited thoughtlessly 'Nature is made for man'[2]. It is apparently clear that slave has no subjection of right or wrong treatment or any expectation of ethical life. This attitude has lead us to scarcity and massive devastation of nature resources.

The aid of science and technology can manage the affairs both ways. The technology is an objective tool. Its use depends on the user. The developed countries or the north block have squeezed the nature on the fiquarative statement of development of planet, but when the question of ethical examination comes on the board, no excuse is there with them. The developing countries are also trying to copy the 'north' but poor fellows do not know that they themselves are getting trapped in a vicious circle of unethical loops for which their coming generation will never give them escape.

"The basic underlying causes of this crisis are mistaken values and attitudes – the attitudes that we can exploit nature without any restriction, that the production of goods is more important than the people who use them, that nature will provide unlimited resources, that we have no obligation to future generation to conserve resources, and that the answer to technology is more technology[3]." Environmentalism has fixed two poles one is human and other is technological management of nature. The ethics of environmentalism has thus a balance of right and wrong just for preservation of present mole of production and consumption. But when at the same time the threats to living species including every living human being, depletion of ozone layers, disruption in food chains and toxication of land, water and air, etc. appear the ethical cry for saving the planet from the group of environmentalists arise, but they themselves know that their own attitudes are responsible for this devastation. This requires the transvaluation of these enshrined values. In fact "the most important part of the environment philosophy is a humane value system which recognizes the ties that bind the commonality of interest between humans and the other species that inhabitate the earth, as well as the interwines fate of all humans in the fact of increasingly severe resource crunches".[4]

"A pollution free environment is regarded as one of the basic rights of the people throughout the work" [5]. but it is not limited to human species, every living being wants on environmental space for its sustenance and beyond that every creature/creation requires its own space. The minimum requirement of space environmental space is an enshrined natural right – the right of animate as well as inanimate world. In fact "environmental space is a right based approach that conceptualizes sustainable development in terms of access for all to a fair share in the limited environmental resources on which healthy quality of life depends. It implies eliminating at least international inequalities in aggregate resource consumption as well as intergenerational one"[6].

Thus here it is around that from the point of view of environmentalism the ethical dictums are – Common good life to all but a subjection of present values, but at the same time the hidden principle of saving the others (including nature and non human living) and also respecting the future generations due gets bleak in eye of environmentalism. The inter-generation equality is superimposed on intra-generation distribution and consumption. The environmental space is a philosophical axiom of ethical structure which should have been in the tenants of environmentslism.

Anthropocentric utilitarian values and economic return values are two schism in environmentalism. But unfortunately both are subjected to present values and thus present generation. These two approaches can form the other hightly important aspects of environmentalist ethics – which can also be chanpioend by shallow ecologists but in fact this is not the true presentation of whole scenario.

The true environmentalism will have to contain the characteristics of deep ecology. Deep ecology is not against the machine or harassing nature but it certainly is against the human greed. The greed in unwanted in deep ecology. A few traces of greed may creep in environmentalism and certainly they creep in and it is advocated by environmentalists in name of preserving the present value system, and this is ethical, but In fact it is not right.

"The deep ecologist believes in biospherical egalitarianism – in principle[7] and it is not biosphere only, it is ecosphere in true sense[8]"

The anthropocentric utilitarianism may be the substitute name for the ethics of environmentalism, but it is not the true situation. The moment of ethics must move from human to non human world and even beyond that in time, space and dimension. The environment space not only exists between generations or two countries "North" and "South" but it also exists as all pervading ethical principle. The reason of this permeability is equality and fair share – the environmental justice. The basic idea behind environmental space in simple : for each individual it is possible to calculate a maximum rate of consumption of environmental resources – a fair share of the maximum available within global limits; while recognising the existence of a minimum determined by need and human dignity.[9]

Thus it is clear that the environmental ethics is an extension of traditional ethics. It requires extension from human world to beyond human world. It encompasses the righteous place and fair share of every constituent on this planet. The ethics which should have cover the all living and non living world up to the limit of cosmos.

The green political theory has analysed both the streams i.e. environmentalism and ecologism. The environmentalists if have belief on machine and technology to reduce the environmental problems of degradation but keeping the present values entact, on the other hand the ecologists advocate the radical change in relation between human and non-human world and in our mode of social and political life[10]. The later view underlines the political theory to be discussed where as the former has stronger element to be included in the discussion.

The environmental ethics, in an empirical way in thus anthropocentric but in an ideal sense it must be ecological and eco-centric, but as it touches the ideal sphere, it looses its independent identity and remains no more, and then it becomes an ontological preposition which examines :

1. Environmental problems

2. Remedies to environmental issues.

3. The flourishing of human and non-human life on earth.

4. Non-exploitation of natural resources except satisfying vital needs in a simple way.

5. Significant change in political and socio-economic vision[11].

It, thus, is clear that examination of world be ideal environmental ethics, decidedly leads to ecophilosophy – the environmental ethics which would provide a justification for our ontological perception[5] of ecosphere.

The additional benefit of defining environmental ethics is to find a platform in shallow and deep ecological stands, because where as the former has specific but blurred picture of ecological ethics the later extends the ethical structure beyond human to even whole existence and finally again going back to environmentalism it is found that "Dominace over nature, natural environment as a resource for humans, natural and economic growth for growing human population, belief in ample resources reserves, high technological progress and solutions, consumerism and national centralized community guide to a working model of environmental ethics but under subjection of deep ecological concerns which have faith in "harmony with nature, intrinsic value of eco-particles eco-sphere equity and equality, sample needs and ultimately a limit to man's endeavor over nature for cause of survival[20].

The environmental ethics is thus a reality of anthropocentric utilitarianism vis-à-vis anthropocentric kant but it sees towards eco-centric ideals in an indirect way by virtue of its own ethical concerns.

REFERENCES

1. Andrew Dobson, Green Political thought, 3rd ed (London : Routlege, 2000) pp. 2-35.

2. SK Chahal, Environment and the Moral life, APH Publishers, New Delhi-1994, p.1.

3. Ibid pp. 2-3.

4. Biswajeet Mukherjee (ed) Environmental Awekening, Sinkha Books, Kolkata, 2005, p.17.

5. Ibid p. 10.

6. Julian Agyeman and others (ed) just lustanibilities, Earth Seen Publication Ltd. London, 2003, p. 25.

7. SK Chahal, op. cit p. 23.

8. Ibid p. 24.

9. Julian Agyeman and others (ed) op cit p. 22.

10. Andrew Dobson, op. cit, pp. 2-35.

11. Bill Devall and George Sessions, Deep Ecology, living as if nature mattered, Pere grine Smith Books, Salt Lake City 1985.

12. SK. Chahal op. cit p. 22.

CHAPTER-8

Green Rights to Survival for Future Generation

Rights are the replica of society and its sanction for the existence of individual. Rights are guarantee for the independent status of man in general and the goodness of the same in particular. H.J. Laski defines "Rights are those condition of social life, without which no man can seek in general to be himself in the best". Rights have social sanctions and enforceability, they are sanctioned keeping in mind the common good along with the liberal theory of individual's freedom. They are claims of individual put before society just for the cause of justice. "Rights appear in every plausible theory of justice and dominate contemporary political rhetoric"[1].

The human rights which have been cherished in the just recent days have the echo of civil, political, economic, social and cultural[2] assertion. These rights are essential for human life. Civil and political rights provide for moral and political order. Such rights invariably include the right to life, equality, political participation and association .... Economic, social and cultural rights are often referred to as second generation rights. These provide substantive stands for an individual's well-being. These second generation rights often have a direct bearing on the human and environment condition. [3]

Rights are the interesting and most cherished subject of political philosophy and their natural existence for cause of individual's claim in society make them more interesting." They are legal guarantee of the means and opportunities for citizens to meet their needs for a fulfilling life. In the same way that the global commons and other common property resources must be fairly allocated and distributed, so corresponding right must be defined and responsibilities assigned to govt. and within society.[4]

As regards the green political thought is concerned, the natural right theory is worth noticing. "A right, must simply, is an entitlement to met or be treated in a particular way ...., rights are 'natural' in that they are invested in human beings by nature or God.[5]

....The 'nature' which is God gifted thus conveys two things, one, in human nature, the appeal of rights against society is there and second, the nature which is around "man" has got its own appeal. Thus the equilibrium between these two appeals form the edifice of green rights in realm of green political theory.

"Nature is a fundamental autumn on which the edifice of green political theory rests, whether in a co-dependence or monistic form"[6].

"The nature which is the sum of what is not the result of human action ironically, this idea has been eco-centric theories" ... It is the wild mountain, or river system without any unnatural 'human presence'. The irony here is that most eco-centric theories appear to work with a monistic metaphysies which consistently denies "dualism", particularly dualism which priortises humans. Humans are envisaged as mere temporary, knots in biospheric met.[7] Thus humans are part of natural order, so it is clear that the claims of humans as right can be claimed but in consonance with nature and its importance. "Arne Naess argues, individual human agents should be considered as Naess knots in the 'biospherical net' and not as a separate factors. This analyses suggests that liberal ideology of humanity has got some natural constraints rather natural obligations. This axiom is just opposite when "Anthropocentric arguments stress that human beings are the sole criterion of value. [6] But the practical or pliant anthropocentrism elaborates" traditional normative theory largely unchanged. However new issues and questions are mapped onto the older normative concerns. Green Political theory thus takes conventional issues of justice, freedom, equality, citizenship or rights and then adds green dimension, emphasizing co-dependency with nature. [9]

Thus the rights in green perspective away between ecocentrism and pliant anthropocentrism. The claim of wild, divine untouched nature stands against the sole nature of human claim. The whole gamut, thus" is about how humanity organizes itself to relative to the nature that sustains it. Thus it encompasses matter of how people deal with the planet and its life, and how they relate to each other through the medium of environment. It impinges on other means of political concern such as those related to poverty, social justice, education, race, the economy, international relations and human rights in as must as what happens is these areas affects our environment (and vice verse) [10]. The existing humanity, has seen and concerned the energy shortage, toxic wastes, air and water pollution, the hazards of nuclear power, biotechnology, species extinction, biodiversity, pesticides, animal rights, wilderness. Biodiversity, pesticides, animal rights, wilderness, protection, climate change, and inequality in the distribution of environmental risks and benefits"[11]. These concerns have awaken the curiousity regarding transference. The treasure of earth is getting exhausted, the exploitation of the same is posing new problems for survivals of human itself and for other non-human natural world too.

The question of transferring the policies of development to the future and for the future leaves an obligation on us. We are to define what we are leaving for that generation which is still to come." It is an obligation to civilisation to continue civilization to pass onto future generations a heritage, natural and cultural, that can be valued and enjoyed with absurdity.[12]

The future generations will have the moral, social, political and economical right. They will have the guarantee for the same also if the democratic values run across centuries, but the environmental rights just for survival of themselves anchor their relevance in contemporary period- the period which they do not belong to.

The coming generations claim their green right to survival from present scenario of pliant anthropocentrism. "In short, it is a concern about the character of the future itself. We want individuals to be happier, but we also want them to have surrounding to be happier about. We want them to have what is worthy of happiness .... What is worth saving is not merely what can be consumed later, it is what we can take pride in and, indeed, love. To protect wilderness and to restore the environment to meet shared ideals is not merely to show respect and concern for future generation but to show respect for ourselves as well.[13]

The relevance of rights in general and the green right to survival in particular becomes much heated when we note that "The population is ever growing, leaving less and less to eat. The air and water are becoming ever more polluted. The planet[15] species are becoming extinct in vast numbers – we kill of more than 40000 each year. The forests are disappearing, fish, stocks are collapsing and the coral reefs are drying.

We are defiling our earth, the fertile top soil in disappearing, we are paving our nature, destroying the wildness, decimating the biosphere, and will end up killing our selves in the process. The world's eco system is breaking down. We are fast approaching the absolute limits viability, and the limits to growth are becoming apparent.[14]

The above scenario depicts the green right of future generation. The sustainability and limits of liberalism are subjected to the right to survival of future generation.

The right to survival comes under category of social or civil rights. This is the most important and vital right of any individual. Be it an utopian eco-centric society or pliant anthropocentric society. The right of survival of human and non-human word is equally in the core part. A ecology can not do without human and a human centred system can't go without nature so the right to survival stands in between.

Now as the future generations are yet to come, they receive the green right to survival and the rights to concern for such right as genetic code form present generation. The transference of parentalism has the intrinsic value for this right. The clean and green environment the nourishing ecology and the normative concepts of political theory make a whole gamut for the future generation.

Humanity's demands on the earth have multiplied over the last half century as our numbers have increased and our income have risen.[15] The anthropocentric version of these demands have evolved the right to survival of future generation for a future.

The increasing population, food production industrialization, pollution and consumption of non renewable natural resources[14] have underlined the nature of exponential growth and this has concern for survival right to future generation. Whether the coming generations may have environmental justice and thus a fundamental right to survival is a big query if the situation of ecological imbalances goes on. The environmental justice frameworks incorporates the principle of the "right" of all individuals to be protected from environmental degradation.[16] The right to survival of future generation gets its strength in the principles of environmental justice[17], when they echo that ethical, balanced and responsible uses of land and renewable resource in the interest of a sustainable planet for humans and other living things, and affirmation the fundamental right to political, economic, cultural and environmental self-determination of all peoples, and the requirement that we as individuals, make personal and consumer chances to consume as little of mother earth's resources and to produce as little waste as possible; and make the conscious decision to challenge and repriotize our life styles to insure the health of the natural world for present and future generaton.[18]

The green right to survival of future generation and the right to concern for the alarming state of affairs appear again when the four big fears come – Natural resources are running out, the population is ever growing, leaving less and less to eat, species are becoming extinct in vast numbers, forests are disappearing and the fish stock are collaposing the planet's air and water are becoming more polluted. Human activity is thus defiling the earth, and humanity may end up killings itself in the process.[19]

Thus the green right to survivalism is not only concerned with the present generation but also the responsibility of present generation and claim of future generation for cause of the existence of this planet and ultimately the whole humanity. This right is the only right which manifests the right of biotic and abiotic world and so in the sense, this is the sole grid of sustainable development; because the later, at the same time feeds back and nourishes the "Green" right in to-to, so the concept of "Green" right to survival reads, "Humanity as the ability to make development sustainable – to ensure that it meet the needs of the present without compromising the ability of future generations to meet their own needs.[20]

REFERENCES

1. Catriona Mc Kinnon, Right : Their basis and limits in Richard concepts. Manchester University Press, Manchester/New York, 2003, p. 16.

2. Tom Bigg, Survival for a small planet, earthseen, 2004, p. 138.

3. Ibid pp. 138-139.

4. Ibid p. 161

5. Andrew Heywood, Political Ideologies, Palgrave Macmillan, New York, 2003, p. 48.

6. Andrew Vincent, Green Political Theory in Richard Bellamy and Andrew Masen (ed), op. cit, p. 187.

7. Op. cit p. 188.

8. Arne Naess, "The Shallow and Deep long range Ecology movement. A summary, inquiry 16(1973) p. 263.

9. Andrew Vincent Ibid p. 183, op. cit. p. 183.

10. John S Dry Jek and David Schlosberg (ed) Debating the Earth, The Environmental Political Reader, Oxford University, Press, 2005, p.1.

11. Op. cit, p.1.

12. Mark Sagoff, The allocation and distribution of resources, in John S Dryzek and David Schlosberg (ed) Debating the Earth Ibid. p. 155.

13. Op. cit pp. 155-156.

14. Tom Burke, Ten Pinches of Salt : A Reply to Bjorn Lomborg in John S Dryzek David Scholosberg (ed) op. cit, p. 80.

15. United Nations, World Population Prospects : The 2002 revision (New York February 2003)

16. Donella H. Meadows and others. The limits to Growth, Universe Books, New York; 1972 pp. 25-44.

17. Robert D Bullard, *www.ejcr.cav.edu/ejinthe21century.htm.*

18. Principles of Environmental Justice, Adopted at the first National People of colour environmental leadership summit, 7/10/1997, Washing ton D.C.]

19. Bjorn Lomborg, The truth about the environment, in John S Dryzeck and David Schlosberg, Ibid, p. 74.

20. World Commission on Environment and Development, our common future, Oxford University Press, 1987, p. 8.

CHAPTER-9

Sustainable Philosophy for Development

The enormous growth of economic sector and changing values of contemporary world towards progress have casted a deep influence on the path of human actions. The inflow of technology and boom in market culture have also changed the pace of globalisation. Now in twenty first Century the economic sector has taken a long jump. But the other associated area have received a proportionate set back. The associating environment, in whose cradle, the economics has got its rejuvinating youth has got a noticeable and even alarming set back. "The current trend in population growth, consumerism, and socio-ecological disorder raises serious concern about the long term sustainability of eco system[1] and it is not the limit, the other aspect in this scenario is that population, food production, industrialization, pollution and consumption of non-renewable, natural resources have got enormous increase. [6] But the same process that have produced a well off society and economic boom have given rise to trends that the planet and its people can not long bear. The failures in management of our environment have been seen. Now there is a more widening gap between rich and poor and more hungry people are there and the pure water, safe and sound homes, sample fuel are going to be scare.[3]

The parameters of development have raised serious question on the survival of humans and other species in this condition. "There are also environmental trends that threated to radically after the planet, that threaten the lives of many species upon it, including the human species" [4].

The economics has another aspect of ecology "it is impossible to separate economic development issues from environmental issues; many forms of development erode the environmental resources upon which they must be based, and environmental degradation can undermine economic development"[5]

The question of a meaningful development, then was raised and the world sustainable development appeared. It appeared first time in the world conservation strategy report (IUCN WWF & UNEP 1980) This report was prepared by world commission on environmental and development headed by Mrs. G.H. Brundtland. In this report the above mentioned concern was not the only concern, the concern for future generation to meet their need out of such environmental affair was also there. The common future of all the citizens and would be citizens was the major concern in this report.

The development on which cost and for what, thrilled the minds of masses. The interdependence of ecological values and economic activities came forward and even for a common global future the interdependence of nations in ecological sphere attracted the minds very sharply. The interdependence of generations was also highlighted.

Sustainable development a development strategy which is thought a balanced approach for ecology and economy in such a state of affairs is a development strategy that manages all assets, natural and human resources as well as physical including financial assets for increasing long term wealth and well being.[6]

The political freedom and civil liberty have enforced the rights or claims of human for fulfilling their needs but at the same time the accountability of this generation to surroundings, non-human world other living world and future generation has subjected this drive with the factor of survival. The loss of life can never be the cost of development. The political movements and socio-economic changes are brought and nurtured for bloom of life not for extinction of the same.

So as has been mentioned earlier that sustainable development is a strategy for "live and let live". Be it the Rio de Janerio Earth Summit (UNCED 1982) or world summit on sustainable development in Johannesburg 2002 with minor differences in approaches the theme became – "finding ways in which limited resources as well as costs and burdens can be more fairly distributed."[7] This need was vigorously felt when it came clearly in light that the globalisation is leading to even more unbalanced development and less sustainable development. It is facilitating market forces and neglecting welfare, social justice and environmental issues.[8]

Sustainable development thus is consistent with the concept of fairness and notion of justice in respect of human world non human world and coming generations.

Now we examine the theme of sustainable philosophy of development. As the name suggest it is not about the path or strategy of development but it is about the basic idea of sustainability. There is very subtle difference between sustainable development and sustainable philosophy of development. In the former concept the characteristics and method of development is central theme while in later the basic idea which makes development is centred.

If we go through the theme of environmentalism we find that here the human needs and the environment have been so managed that no harm be reached. The sustainable development finds its place in environmentalism "Environmentalism argues for a managerial approach to environmental problem, secure in the belief that they can be solved without fundamental change in present values or pattern of production and consumption[9]

The sustainable philosophy of development underlines the worth and dimension of development. Whether the manifesting development is development or otherwise. The human who is defining development will survive or not just because of the development. In which direction and dimension the development should extend and for what because it is philosophy of development so along with human other life forces and non-life factors have also claim for development. The development of human can't be the sole criteria. The holistic impact of idea for development of all is the theme of sustainable philosophy of development. This guides to ecologism where the mutual development exists. "Ecologism holds that a sustainable and fulfilling existence presupposes radical change in our relationship with the non human natural world, and in our mode of social and political life.[10] Thus besides having a holistic concept of development of all it also underlines that it is not a simply economic or political or social development. It is a wider term which incorporates every development – the development of every aspect. So in nut shell it is holistic development of holistic world. This also encompasses moral asthetical and spiritual development. Ecologism incorporates the holistic world, undoubtedly. The intrinsic value of whole creation and its culmination is the sustainable philosophy in perspectives of ecology. This philosophy consits of the ecological integrity and prosperity and community livability.

The sustainable philosophy also points out the working philosophy behind eco-centric and anthropocentric models of society. The deep ecology puts the nature in centre but the shallow ecology keeps the man in center. The former is ecologically conscious whereas the later talks about the personal development. The former is radically holistic whereas the later in reluctantly holistic[11] and these two wings come under umbrella of ecology. The ecology is an ideology which encompasses the central theme of ecological balance, holism, sustainability; environmental ethics and self actualisation.[12]

Thus the sustainable philosophy wants to develop all the factors from all the aspects. There is no distinction or management but we will have to agree that a fair adjustment among the actions and claims of human and non human world well have to be reached. This we read from distinction between deep ecologism and shallow ecololgims.

The different concepts in political theory which have come from various disciplines like sociology, psychology, economics philosophy, anthropology, theology and journalism[13] draw our attention when we talk about sustainable philosophy of development. Political theory has borrowed various concepts. The patrimony of political science is full of boomed concepts[14]. The concept of sustainable philosophy of development, can be added on the background of ecological understandings but to what extent should ecological factors be included in the parameters of economic; social and political development.[15] The answer may be fairness, placement and the allotment of worthiness for all the players for the cause of survivalism and removing arbitrariness in allocations and respect. The political economy talks about the redistribution of debt.[15] The ecologically oriented political theory talks about the intra and inter-generation distribution of allocation of sources. Exhausting life support sources, exhausting energy sources, depletion of biosphere, deforestation, acid rain, global climate change, expansion of deserts, infertitly of soil, pollution, global dumping grounds, health hazards etc. have loaded the coming generation with ecological debts. This state is alarming the present generation also. So the sustainable philosophy takes into account the sense of justice in maintenance of resources and their allocation.

The sustainable philosophy of development encompasses the basic canons of environmental justice in it. The respect to sacredness of mother earth, ecological unity and the interdependence of all species and the right to be free from ecological destruction, mutual respect. Justice for all peoples, free form any form of discrimination or bias, ethical balanced and responsible uses of land and renewable resources in the interest of a sustainable planet for human and other living things, keeping the pollution, radiation and toxility away and the fundamental right to political, economic, culture and environmental self-determination of all people[16] are the trigger points of sustainable philosophy of development; and beyond this the existence and claim of non living word and non human living are equally have an appeal for their self-determination. The values attached to the non living and non-human world have been functional at human level, and the same has been chief spokesman since the dawn of creation but in no way they can be given superiority. The sustainable philosophy of development talks about eco justice. A society which aimed to reduce consequential remoteness and open eco justice issue to effective political action would need, among other things, to be participatory and communicative, and it would need to be society of substantial equality and democratic culture ....an ecological rational society would need to be more ambitious in this direction then any society we know now, but this may ultimately be the condition of our survival. [17]

The condition of sustainable development manages for environmentalism where the human existence is accepted in underline and the production and consumption can be adjusted technologically or decentralizing the socio-politico- economic forces and environmental problems imbalances etc. can be managed accordingly but the induction of sustainable philosophy for development pushes the entitlement for environment also in four ways first, the human entitlement to productive, diverse and fully functioning bio sphere, second, sharing the critical natural resources for balanced development , thus, the citizen entitlement to natural assets required to meet basic need and fourth, the inter-generational entitlement to each of the above three[18]

The sustainable development issue raises the question about the basic attitude of human to live with limited resources, respect for bio diversity, visualization of future, socio-economic feasibility of new technologies and how to achieve the targets set in them. [19] and leaves the question unanswered but the sustainable philosophy gudies the path that "Meeting the needs of the present generation while ensuring that future generation inherit the capacity to meet there needs in a comprehensive development objective. No one is excluded ..... And that includes maintaining the unique environmental services and goods (or critical natural capital), such as stable climate or diverse eco system and their product flows, which all people require and will continue to require. To sustain critical capital, overall pollution loads, and resources take must be kept within the tolerance limits of biosphere and its primary ecosystems. Otherwise they will collapse." [20]

The sustainable political philosophy, for development, by the way echoes "a sustainable political ecology, whether at global, national or local levels, has to marry social justice and equity with economic development and environmental protection in order to enable people to meet needs fairly" [21] and thus up keeping and continuing the revolve of nature in its natural stage.

At the same time it is noted that "Ecological sustainability is replacing or integrating economic considerations with ecological considerations... The primary value of ecological sustainability.... Without the integration of ecological sustainability with economic sustainability, there is no foundation for human life or for other species. .... A sustainable economy integrated with ecological and social consideration is the goal of our common future[22].

REFERENCES

1. DK Verma, sustainable development, M.P. Journal of Social Science, Vol. 4, no. 1999.

2. Donella H Meadows, and others, The nature of exponential growth in John S Dryzeck David Schlosbug Oxford University Press, 2005, p.11.

3. World Commission on Environment and Development 1987 in John S Dryzek (ed) op. cit p. 260.

4. Ibid p. 260.

5. Ibid. p. 260.

6. D.K. Verma op. cit p. 41.

7. Tom Bijg (ed) survival for a small planet the Sustainable Development Agenda, Earth Scan, London, 2004, p. 19.

8. Ibid p. 26.

9. Andrew Dobson Environmentalism and Ecologism, in Alan Finlayson (ed) Contemporary Political thought Edinburg University Press, 2003, p. 364.

10. Andrew Dobson, Green Political Thought, Rootledge London 2000, p. 2.

11. Andrew Heywood, Political Ideologien, Palgrave Macmillan, 2003, p. 272.

12. Ibid, pp. 269-283.

13. Robert Egodin and Hans Dieter Klinge mann, A new Handbook of Political Science, Oxford University Press, 1998, p. 103.

14. Ibid 102.

15. Ibid 119.

16. Principles of Environmental Justice, adopted at the first National people of colour environmental leadership summit, 07/10/1991, Washington D.C.

17. Val Plumwood, Inequality, Ecojustice, and Ecological Rationality in John S Dryzek and David Schlosbeg op. cit p. 629.

18. Tom B 199. op. cit 160-161.

19. DK Verma op. cit pp. 56-57

20. Tom Bigg of. Cit p. 158.

21. Tom Begg op. cit p. 159.

22. V.K. Natraj and others (ed) Gandhian alternative, concept Publishing company, New Delhi, 2005, pp. 148-149

CHAPTER-10

ENVIRONMENTAL JUSTICE : THE CORE ISSUE

Etymologically the term environment sounds surroundings. "It is an English word formed by two words, i.e. 'environ' and ment, which means 'encircle' or all round"[1] means everything be it living or non living comprises of environment. Air, water, earth, fire, sky, vegetation, wild life, biotics, human beings. Inhabitation, the stage of development, technological instruments and the whole existence, thus forms a complete picture of environment.

The missing or extension of political philosophy or the traditional concept of web of social relationship to realm of environmental whole forms the miraculous prologue of environmental justice. The interchanging dynamics of human and all non-human cores the central platform of justice in relation to environment. In fact, the concept of justice is a social concept[3] which denotes a harmonious reasoned, rationale social engineering for a better life. It is an understood philosophy that justice, thus denotes an arrangement of associative and distributive pattern for a fair society.

Today when the capitalism, globalisation, expansion of digital revolution, extension of information and liberalization are taking place in its full pace and the growing seals of world ecomies have established the cause of environmental justice firmly[4]. The ruthless exploitation of nature and abiotic biotic and non human world for satisfying the greeds and needs of human has disruted the natural balance of whole existence. Infact a harmonious and respectful relationship between human, and non-human world is the core concept of environmental justice. If is apparently clear that for such socio-biology an ethical force works behind. This ethical force which works in name of environmental management, environmental protection or environmental evaluation or any other name, ultimately is the synonymous of environmental justice. The interacting actions thus be in between human and human or human and the whole remaining underlines the concern of environmental justice[5]. Till now the concept of justice was playing in cradle of society, but now the environmental justice plays with society at one hand and rest whole on the other.

The principles of environmental justice enshrined in first national people of colour environmental leadership summit[6] reads under summations as under:

1. Mother status has been given to earth. It is sacred. All species are interdependent and have the right to be free from ecological destruction.

2. Public policy be based on mutual respect and justice for all people. (This affirmation rejects any preference to any sphere "Developed" privilleges or "underdeveloped") burden.

3. Reasoned, rational, ethical balanced and responsible use of natural resource for sustainable life of this planet. The life comprises of all living beings.

4. Pure, clean, safe, healthy and harmonious enrichment is the right of every citizen of this planet.

5. The environmental consciousness must be passed unto coming consecutive generations.

These principles mention the unity and entity of planet as sacred, and further that every constituent of this sacred mother earth is inter dependent. These principles do not provide eccentricity either to anthropocentrism or ecocentrism, as the two thoughts run parallel in green political theory, but they emphasise the one organism theory. The core of idea of environmental justice the interinse value of existence – that itself resides or pervades the whole existence, at the same time, "the question of environmental justice not anchored is a debate about whether or not decision makers should think with risk management. The framework seeks to prevent environmental threats before they occur.[7]

Thus it is clear that a warning before ecological degradation may take place is equally a principle of environmental justice framework seeks to prevent environmental hazards before they take place through harmful practices in industrialization, land use, globalisation, and other developmental works.

Environmental justice advocates challenge the current environmental protection apparatus and offer their own framework for addressing environmental inequalities, disperate impact, and unequal protection[8] and very silently but emphatically this framework brings to the surface the ethical and political questions of "who gets what, why and how much".[9] Exploitation of land, environment, nature in general and of people in particular is another concern of environmental justice. This is not allowed in realm of environmental just order. It has been observed that deprived class or poor people, generally work and live in most harmful conditions. The poor nations become dumping ground of toxics. This all happens just because some are powerful and out of their unequal rights of privileges, environmental justice condemns it.

The concept of social justice may accept the weighted distribution of social fruits according to merits but one step ahead the environmental justice recommends the bare right of living just for cause of life, so that on this equal platform the distribution of social fruits may be reengineered for the whole existence.

Environmental justice puts at one side different mind and their merits and on the other hand the life the nature as a whole organ. It hints the ethical actions towards sacredness of earth by altering the mind as "our minds are the products of interplay between our external environments and our genetic inheritance. The environment in which we live is also influenced by nature and the people who live in it. It is harmony and balance between the two which can lead to the ultimate self activation we strive for.[10]

Environmental justice, thus is a guarantee for our common heritage. It never supports any extrimism be it in name of development or white man's burden, or north dominance over south. The common heritage believes in equitable distribution but without any exploitation, and a fearless democratic world order and also assures the future generation the inheritance of pure, safe, clean and faithful environment.[11] This concept of environmental justice runs in time. The present as well as future generation insurance is guarded through high importance. Thus at one look where as this concept becomes an end at the other look it becomes a look for the cause of itself – such a miraculour wonderful idea of real sustenance of life. Perhaps this is the same life which Aristitle says a good life[12] and this life extends beyond human to all non-humans. Environmental justice thus appears a very important concept of socio-biological engineering. A fair distribution not among human world but beyond human world for the cause of sustenance of nature as a whole organ.

Environmental justice has also cast its deep impact on international agendas and on international law. The Stockholme declaration 1972, Helinski Conference 1974. London Conference 1975, Viena Conference 1985, Montreal Protocal 1987, World Commission on Environment and Development 1987 Rio conference 1992, Kyoto conference 1997, Johannesburg Conference 2002, etc., all these international platforms have raised the question of lessening the pollution and exploitation of nature, rationalized use of land, balancing the atmosphere, curbing the hazardous toxins rays and industrial wastes, safety isues of humanity, inequality of social opportunities of life, poverty alleviation for a good life a environmental just life.

Philosophically the modalities can be rethought in different ways to equalize the environmental protection and its sustenance, but at the same time, the core issue of biological continuity of species in general and human in particular along with the varied non-human world is the central platform of environmental justice.

REFERENCES

1. Biswajeet Mukherjee, Rahul Ray (ed), Environmental Awakening, Sikha Books, Kolkata, 2005, p. 259.

2. Oxford Dictionary (Indian Edition) 1994, p. 403.

3. Deninger W Problems in Social & Political thought, A philosophical introduction, McMillan, New Delhi, 1965, p. 221.

4. Saxena Rajeev, Environmental Justice : An insight, MP Social Science Research Journal, Ujjain –1, June 2002, 1.

5. Ibid, p. 71.

6. Principles of Environmental Justice, adopted at the first National People of Colour Environmental Leadership Summit, Washington D.C., Oct., 1991.

7. Robert D Bullard, Race and Environmental Justice in the US Tale Journal of International law, 18 (winter 1993) 319-35; Robert D Bullard The Threat of Environmental Recism, Natural Resources and Environment, 7 (Winter 1993), 23-6, 55-6.

8. John S.Dryzek, David Schlosberg, Debating the Earth, Oxford University Press, 2005, p. 433.

9. Ibid. p. 433.

10. Nandini Chakraborty, The mind and the environment a continuous inturplay, in Biswajeet Mukerjee (ed) Environmental Awakening, Sikha Books, Kolkata, 2005, p. 179.

11. Das RC, Barel JK and Others, The environmental divide, APH Publishing Corporation, Delhi-199, pp. 182-183.

12. Foster MB, Philosophy on Political thought, George and Harap Co, Ltd. 1961, p. 124.

CHAPTER-11

ECO INJUSTICE : THE CRISIS OF COMMITMENT

The normative concepts of political philosophy like rights, liberty, faternity, equality, harmony, solidarity, obligations, law, etc., have been co joined on a single platform and that is known as justice What is justice, can't be said in a water tight compartment style. It is a moving concept.[1] The principles of justice are regarded as formulating restrictions as to how practices may define position and offices and assign, there to powers and liabilities rights, and duties...., the meaning of the concept varies according to whether it is applied to practices, particular action or persons"[1] Thus in a typical society the justice in disposition as per rooms of set standards, but ultimately justice is fairness in substance. This fairness is subjected to two standards. "First, each person participating in a practice, or affected by it, has an equal right to the most extensive liberty compatible with a like liberty for all and second inequalities are arbitrary unless it is reasonable to expect that they will work out for everyone's advantage.

These principles express justice as a complex of three ideas : liberty, equality and reward for services contributing to the common good.[2]

Justice thus, underlines an assertion of self but subjected to common ethical standard, and in this light the canon of distributive justice comes in light. The complete principle of distributive justice would say simply that a distribution is just if everyone is entitled to the holdings they posses under the distribution."[3]

The ecological players in the light of justice separate two worlds "Biotic world" and "abiotic world". Both these worlds interact and the interaction within them go simultaneously, so the perception of eco-justice adopts a complex mechanism. – the relation among beings (humans) and non humans and the relation between humans and non-humans. The nature which is the central player of green political theory conveys that, the concept of nature implies a source or principle of action that makes something behave in a certain way .... It is an idea familiar from Greek philosophers to the present. However, there is second conceptual sense of nature that refers to the sum total of things and events. The sense of the 'sum total' can also imply two different ideas : first it can signify those things which are distinct from human action; intention or artifice. Another way of putting this is nature refers to those things which are driven by patterns of causation distinct from human action.[4]

The nature has its own identity and the humans are biospherical knot and the nature has got the natural justice for its own existence. The nature compress of the humans the nature itself is a value and virtue simultaneously. The biotic and abiotic world excluding human, thus has its own unique signature and they have got their right to be themselves by virtue of their being natural. This is the essence of eccentric political thought. The natural elements be treated in just manner and as per distributive justice they must hold what belongs to them and arbitrarily they may not be separated from their dues. Though the term eco-centric or bio-centric implies that intrinsic value is located in nature and can connote an absence of regard for human interests, essentially shedding one centrism in favour of another. Yet the variety of human interests in nature to be sheltered under eco-centric umbrella. The justice and fair disposition thus rolls between the two values where both the spheres are under umbrella of nature. The ecological justice thus is fairness among reactions and moral disposition of human and non-human word. "Such a moral stance implies bio-centric equality, the principle that all organism and entities in the ecosphere are of equal moral worth, each of them being part of an interrelated whole.[5]" The equal rights to bloom[6] defines the ethical principle of eco justice.

The rights and liberties of humans be defined and limited in such a manner that the dues and claims of non-human world may be protected. The bio-centric and anthropocentric pattern of just disposition may be a crisis if both are weighted separately without respecting the other.

Vandana Shiva[7] writes that "with the destruction of forests, water and land, we are losing our life support system. This destruction is taking place in the name of the development and progress, but there must be something seriously wrong with a concept of progress that threatens survival itself.

This notion elaborates that the domain of progress induced by human should not destruct the domain of survival support system, and on this anvil the respect and regard to other non-human elements should exist. The fairness and ethical equilibrium should be there.

Justice is not a blue print of wrongs and rights. In fact, these are tools for core concept of justice according to culture, but the ecological justice springs op for the cause of survival and this is based on fair play. "Rawl's theory of justice is precisely the notion of justice as reciprocity that is embodied in the principle of fair play. According to Rawls, a society is a scheme of social cooperation, and from this fact we can generate, via, the notion of fair play, principles of justice"[8]

The liberty rights and equality rest ultimately on the ground of fair play – the ground which presupposes the regard for survival. The green justice rests on fair play and due regards to both the world – biotic as well as abiotic world. The eco-anarchism underlines the background of green justice when Bookchin suggests that there is a clear correspondence between the uncoercive authority and ecology. This blend is there is social ecology. This concept believes that ecological balance is the surest foundation for social stability. Anarchist believe in stateless society in which social solidarity mutual respect among human beings exists and the ecologists believe in harmony within nature.[9] Thus both the concepts hint towards a finer blend of human and nature more precisely the balancing of nature whose integral part in human also. A respect for diversity, balance and harmony, thus underlines the essential principle of green justice.

For induring the principles of justice in human world the political authority reinforces the sovereignty of nationalism. But ecologically rational society is unlikely to be found where the kinds of political structures and culture necessary for human justice and communicativeness are also lacking[10]. Diversity enhances the potentialities of survival, the chances of new modes of life, and the richness of forms. And the so called struggle of life, and survival of the fittest should be interpretated in the sense of ability to coexist and cooperate in complex relationship, rather than ability to kill, exploit and suppress. "Live and let line" is a more powerful ecological principle then" Either you or me"[11]. The sense of fairness for equitable distribution by putting the players of ecology at appropriate creates the appropriate sense of ecological justice. The moral parameter of human demands that other organisms or non-organism be treated with care and consideration, but this consideration from human view may treat other non human world with not giving moral regard. [12] And if this happens this will be ecological injustice. As a member of mother earth planet then we must join together to bring forth a sustainable global ecological society founded on respect for nature, universal human rights, economic justice and a culture of peace. Towards this end it is imperative that we the people of earth. Declare our responsibility to one another to the greater community of life, and to future generation[13].

The ecological justice not only takes into consideration the present generation but also talks about future ecology and generation. The idea of fairness, mutuality and regard for the cause of survival extends the intrinsic value of justice and when the nature which comprises of human world as an integral part of itself besides other non human world is given the prior value than any thing else, the evolution of ecological justice takes place.

The tools of distribution and retribution play with in the umbrella of ecology. These two tools must be extended beyond the human world. The moral category of distributive justice to non-human life form and covering of all forms of life on the planent echoes the presupposition of ecological justice. [14]

At this point it should be noted that the ecological justice is a bit different then environmental justice. "Environmentalism argues for a managerial approach to environmental problem secure in belief that they can be solved without fundamental change in present values or patterns of production and consumption, and ecologism holds that a sustainable and fulfilling existence presupposes radical change in our relationship with the non-human natural world, and in our mode of social and political life" [15] The environmental justice presents a bit change in existing values of humans but more or less the concentration is on anthropocentric axis – the human word, the morality is not to non-human world it is central to human needs, rights equality and liberty; the concern for moral worthiness to non human world is not here. The ecological justice, at the same time is a demand for moral order of whole of the creation. The fairness out of moral action comes in both the world and the interaction starts. The shift here is not only on the human needs and desires but towards a largest circle which comprises the needs, claims, existence of non-human world and human world as well. Ecologism's through going naturalism rests on the belief that human being are natural .... This is a natural world in which interdependence is given priority over competition and in which equality comes before hierarchy. Nature[16] "in this view, no part of the natural world is independent and therefore no part can claim to superiority" [17].

So ecological justice is a moral distributive plan for each identity biotic or abiotic, where mutual respect and fairness of affairs exist and the existence of every creation is expected and finally the superiority of natural order exist not the human centred order. In a liberal democratic society such type of axiom can best be achieved by a ecologically political rational authority, the very authority who decides which decides who gets what when, why, where and for whom.

REFERENCES

1. John Rawls, Justice as fairness, in Robert E Goodin and Philip Pettit (ed) contemporary political philosophy (ed), Blackwell Publishing, 2006, p. 185.

2. John Rawls, op. cit pp. 186.

3.. Robert Nuzick, distributive justice in Robert & Gooding and Philip Pettil (ed) op. cit, p. 201.

4. Andrew Vincent, Green Political Theory in Richard Bellanny and Andrew Mason (ed) Political concepts. Manchester University Press. Manchester, New York, 2003, p. 187.

5. Andrw Haywood, Political Ideologies, Palgrave maemillen, New York, 2003, p. 281.

6. Arne Naess, community and life style ruthine of an ecolosophy, Cambridge University Press, 1989.

7. Vandana Shiva Development; Ecology and women, in Alan Fin layson (ed) Ccontemporary Polytical Thought, Edinburgh University Press, p. 2003, p. 422.

8. Brian Barry, Humanity and Justice in global perspective, in Robert E Goodin and Philip Pehit (ed). op. cit. p. 727.

9. Andren Heywood, op. cit p. 287.

10. Val Plumwood, Inequality, Ecojustice, and ecological rationality in Josh S Dryzek and David Schlosberg (ed) op. cit p. 614.

11. Arne Naess, The Shallow and Deep long range ecology movemet : A summary (in) John S Dryzek and David Schlosberg (ed)op. cit 344.

12. Brian Baxter, A theory of ecological justice, Routledge, New York, 2004, p. 189.

13. The Earth Charter, *www.earthcharterinaction.org.*

14. Brain Boxter, A theory of ecological justice, op. cit. p. 8.

15. Andrew Dobson, Environmentalism and Ecolism, 2000 in Alan Finlayson, (ed) contemporary political thought Edinburg University Press, 2003, pp. 364-365.

16. Ibid. p. 379.

17. Ibid. p. 380

CHAPTER-12

ECO FEMINISM : AN INTRODUCTORY FRAMEWORK

In twentieth century, the idea of dominance of sex on political philosophy emerged. It was thought that all major thought streams have been designed by males and for males and, there is no place of other sex. Even in name "individual" a male dominance has always been kept in mind in annals of history. In this state of affairs how a just and fair position of both the sexes can be thought of Feminists who advocate for female rights and their equality in every affair and their existence, "have long recognized that political philosophy has been constructed with male heads of households in mind so that the universal right of men were just that did not include women."[1]

The appeal of females position and its assertion is one aspect in contemporary political thought at the other women became the embodiment of biological function the image of nature the subjugation of which constituted that civilization's title to fame. For millemia men dreamed of acquiring absolute mastery over nature, of converting the cosmos into one immense hunting ground. It was to this that the idea of man was geared in a male – dominated society. This was the significance of reason, his proudest boast"[2]

Thus the nature was thought to be vergin and sacred and life female or vice verse. In green political thought the females are believed to be closer to nature than men. That gives women a particular stake in ending the domination of nature.[3]

The domination and exploitation of nature by human race has been put as cruel exploitation of the nature that is female. History has presented with the examples when the dominance aggression and suppression of females have taken place vehementy. The females are thought to produce or create with sacredness and compassion. Females are weighed with harmony and cooperation. They are nurturing entities. Females have been put in rules of the mother – the life giver, the wife, the creator, the sister – the compassion giver or grand mother – the shower of love and affection. In the same way the nature is thought; it is equally having all those properties which a female has. Ecological science concerns itself with the interrelationships of all forms of life. It aims to harmonize nature, human and non-human. It is an integrative science in an age of fragmentation and specialization of knowledge. It is also a critical science which grounds and necessitates a critique of our existing society. It is reconstructive science, in that it suggests directions for reconstructing human society in harmony with the natural environment.[3]

Feminism does not, at the same time advocate for hatred and brutality rather it advocates the existence of all with love and affection. Being in tune of ecological science as narrated above the ecological feminism or eco feminism pushes social ecology to understand the necessary connection between ecology and feminism so that social ecology can reach its own avowed goal of creation a free and ecological way of life." [4]

In philosophical theme the basic principles of ecofeminism are

1. Females are closer to nature.

2. Subjugation of nature is alike suppression of female

3. Hierarchy and seperatism are not the themes but a web of harmonious bio social relationship with due respect every constituent of nature stands there.

4. Politics and culture of ecofemism do not support tyranny or cruelty. Ecofeminism stands for love and mutual existence and so the actions.

5. Survival of all species and non human world including human itself along with non living natural creations is the basic theme of ecofeminism.

And this whole theme of ecofeminism stands need a decentralized global movement founded on common interests but celebrating diversity and opposing all forms of domination and violence. Potentially, ecofiminism is such a movement[5]. The strategy for dealing with this issue was originally canvassed by the social ecologists Murray Bookchin. It involves a subtle blending between nature as distinct and nature as integral to us[6].

Ecofeminism is integrative. The ethics of ecofeminism is thus respect for nature and not preservation of nature just due to its usefulness but because of its inherent worth[7].

At the same time be capitalistic or socialistic or liberal democratic political economies the environmentalist consciousness underlines the preservance of life. By use of technology the economics can be developed and certainly it is mandate. But no technology wants destruction of nature. Control of nature is different than destruction of nature. Anthroprocentric eco feminism is advocate of controlling the nature (but there is preservence of nature as it had its own worth – intrinsic worth) and ecocentric ecofeminism is staunch movement for natural way of life where man is allowed to lie with non-supressive technology but man is not allowed to disturb nature any more. In this school the man's life is there just as a knot in a big net. "Ecofeminism supports utopian views of harmonious, diverse decentralized communities, using only those technologies based on ecological principles, as the only practical solution for the continuation of life on earth"[8]

Industrialisation, globalisation and liberalization can not be rolled back, be there millions of moment against the same but a roll out of ecofeminism can certainly take place simohaneously which teaches non-violence and responsibilities for coming generation. A military culture and violence against any caste, sex, race, creed or human – non human world is never allowed in eco feminism. The intrinsic values of nature, at the same time, go by this dictate. The ecofeminists thus revolt against any structure or institution that is against interest of life, but as an indepent existence, ecofemism does not offend the, bio-sociology, it does not usurp the power in hands of women alike patriarchal society rather it advocates for no power at all and the human beings will have to be treated as human beings; and not as male or female.[9]

Thus a relationship with nature in non dominating way is "ecofeminism and with all the due regards to anthropocentrism, ecofenism is not concerned much with right, obligations, ownership and responsibilities, but with care, appropriate reciprocity and kinship"[10]. These traits, ultimately are not going to disintegrate the whole ecology or disrupt the living environment.

"Nature" has been equalized with female and "culture" with males. The ecofeminism recognizes the triumph of nature of nature; as since dawn of human history the suppression of nature and females have taken place. Feminist environmentalism is thus a parelled mandate which treats nature with harmony love, reciprocity, natural affirnity and non-tyranny. These are the qualities of any female. "Environmentalism argues for a managerial approach to environment with aid of technology and without disrupting present values of production and consumption"[11]. Now the female environmentalism aids the female qualities and psychology to managerial approach to environment.

Ecofemism also opposes overpopulation with conquence and distruction of natural resources and unfortunately for both causes modes are responsible, as the dominating characteristic are naturally with male. Ecofemism is, thus diagnolly in opposition to androcentrism – the dominance of male in nature and in females[12]. and consequentially the females get automatically drifted towards the nature. This repulsion generates attraction.

It is also note worthy that the subtle traits of eco-feminism also generate the environmental ethics which are but natural female by characteristics. Not to harm any one, cooperation, life to all and peaceful movements and finally a soft appeal of 'live and let live' are, certainly the traits of ecofeministic ethics.

The subdivisions of ecofeminism in shape of social ecofeminists (which advocates materialistic methods to analyse present capitalistic economy). Cultural efoceminists (which advocates associative ways to use the present environment for need fulfilling or social ecofeminists which talks about the gender equality and thus a new make –female relationship with nature and transformation of society for sustainability) all talk about the soft and gentle handling of nature for the life of all.

The ecofeminism thus is not a "theory" or a new "ism" but a newer approach to perceive the nature and its relation with 'man'. The thrust approach which has philosophy of "live and let live" at all in its complete unfoldment with compassion.

REFERENCES

1. Karte Nash, Feminism, in Alan Finlayson, Contemporary Political Thoughts Edinburg University Press, 2003, p. 312.

2. Horkheimer and Adorno, Dialectic of Enlightenment, New York, Seabury Press, 1972, p. 248.

3. Yenstra king, towards an ecological feminism and a feminist ecology, in John S Dryzek, David Schlosburn, Oxford University Press 2005, p. 399.

4. Ibid. p. 399.

5. Ibid p. 400.

6. Ibid p.400.

7. Richard Bellaning and Andrew Mason (ed), Political concepts Manchester University Press, New York, 2003, p.190.

8. Chahal SK, Environment and the Moral life, APH New Delhi, 1994, p. 82.

9. Ynestra King, op. cit p. 404.

10. D. Eauborne P., The Time for Ecofemenism, in C. Merchant (ed), Key concept in critical theory; Ecology, Rawat Publications, New Delhi, 1996, p. 174-197.

11. SK Agarwal, Green Management, APH Publishing Corporation, New Delhi-2005, p. 250.

12. Andrew Dowson, Green Political Thought, Routledge; London, 2000, pp. 2-35.

13. SK Agarwal, op. cit p. 254.

CHAPTER-13

THE PHILOSOPHICAL VALUE OF ECO-DEMOCRACY

The values enshrined in democratic theory manifest the clarion call of self actualized liberaties with reason of good inhabitation on earth. Though the traditional political theory has talked a much about traditional concepts of political philosophy while dealing with democratic theory, yet the values of life and preservation have thrived in it.

In eco-democracy the horizon of thought goes beyond the democracy among humans, it also touches the beyond human echelons of rights and duties, existence and appeals of peri human world. The Green political theory has dichotomy of anthropocentric and ecocentric models but ultimate question of presurvance of value of life rests with. The responsibilities, the liberties and the transcendence from human and non human communities for cause of flourishment of life is the sole guiding value of eco-democracy, because this pattern of democracy moves beyond humans and incorporates the whole planet and its constituents.

The claim of "nature" and "technology" and "needs" and "deeds" have ushered the democratic theory on ecological canvass for cause of life itself. In this context ecocentrism is a louding call to the theme of, if not wrong literally, back to nature. And this word has got a massive momentum while deciding the horizons to move for civilizations. Opposing to this move some are of the school of anthropocentrism that has belief in control over nature for cause of needs and deeds of human species. Green movements or 'deep', 'shallow' 'dry' greens etc. have different views of harmony between human species and non-human whole world, but in actuality, there is a hidden anxiety of balancing the both. There is now not a factionalism but "a number of 'second wave' green political theorists are now inclined to reject the anthropocentric/ecocentric distinction as unhelpful and have sought instead to emphasize the humanist credentials of green politics and redirect attention towards questions of human ecological stewardship, ecological virtue, and what might be the legitimate use (or illegitimate abuse) of natural resources and ecosystems"[1].

The concept of sustainable development reels the same concern. The political philosophy of this debate has taken, shape of the sustainable development and our common future. Whereas these ideas are here, some coined that this is an untruth full thinking the focus on convergence among environmental and social concern vis-à-vis democratic theory[2] gained the momentum.

The political theory and Green political theory (though the later is an offshoot of main rail of political theory), infact a confluence for a common stream of "Our common inheritable future". The voice and appeals of not only non-human world of present times but, the calls and representation of all the constituents of future generation, thus can be the eco-democratic theory of the coming times. The stress of maintaining democratic expectations, at the same time lie on the non-anthropocentric thinkers, who are to open the gates for moral interventions of human existence along with necessity of non-human element[3]. 'Back to nature' theme is to be read with "morale" of human element. "The internal balance within ecosystem entails a web-link interplay between many species and their inanimate environment"[4] and that is the beginning of socio-econo-political ecology.

The liberal democratic traits of traditional political theory get coloured with eco-democracy of future. The control of reason, rationale, merit and justice takes the call for survival of future world. So the "democracy" adopts the new shape of greater responsibilities. Even the traditional concept of sovereignty and supremacy of state gets controlled with the newly arising green movements. The philosophical quality of democratic value that is to sustain, support and supply of "good life" has got the endorsement again. And wonderfully the root to achieve them is not an interplay in human world only but in whole creation where the rights of animals and vegetation and even the natural unloving objects have been counted. This has been pointed out more emphatically by Arnae Naess and others – If we try to exist separate from nature our self – realization will be blocked. Thus if we want to exist fully we cannot destroy parts of nature"

The eco-democratic philosophy will thus require the changes not only in social, political or economic level but a whole look at affairs. The articulation of bio-continuity, survival of life support system, science, philosophy and religion are some other platforms where the ultimate question of eco-democracy lies.[6] Eco democracy thus has internal and intrinsic value of nature.

The eco-democracy has also raised the voice for animal liberation and when back to nature theme of ecocentrists is thought on the natural nonliving world liberation also takes place.

Now on the practical aspect it may be a riddle to gauge the liberation of non-human world. Just to save any animal or vegetation from brutal killing can't be the sole guiding factor. Up to what actualization they want to thrive, this we humans can't vouche. Because Zoe have trans human values, therefore we can sense them but only in our perception. Because the non-human world has no socio-political pressure group, therefore at the best, to avoid them from extinction and give them suitable life condition can be the dictate of eco-democratic theme in this regard. It is also to be noted that these dicates, if are purported to put human needs behind then, they are defective, infact, just for the cause of their existence, the non-human world must survive and grow. We can have many reasons for our conservational efforts – not all of which are parasitic on our own living standards (or need fulfillment) and some of which turn precisely on our sense of values and or acknowledgement of our fiduciary responsibility.[7]

The democracy is saluted for call of development and in name of anthropocentrism, the claim regarding the same are cherished. In fact it should be but when the situation is vouched from eco-spectacles we find that tilt of development must not destroy non-humans on the cause of development "Development is fundamentally an empowering process, and this power can be used to preserve and enrich the environment, and not only to decimate it.[8]" Preservation has its own intrinsic value which humans can vouche. In the same way the empowerment of weaker gender and species, animals, microbial world can be thought of.

The affinity of democratic value as enshrined in "Green Marxism" can be very best implemented by allowing the sustainable philosophy for development. "Green Capitalism", "Green Socialism", "Green - authoritarianism" and "Greenism" of different ideologies ultimately lead to the ultimate question of space of life – a qualitative life for humans of present and future and a qualitative as well as quantitative life rights and respect commanded by non-human world.

REFERENCES

1. Tim Hayward, Political Theory, and Ecological Values, St. Martin's Press, New York, 1998, p. 5-6.

2. Kate Sopor, What is Nature, (Oxford, Black Well 1995)

3. Robyn Eckersley, Ecocentric discourses, Tamkang review vol XXXIV, No. 3-4, Spring summer 2004, pp. 155-86.

4. Biswajeet Mukherjee, Environmental Awakening, Sikha Books, Kolkata Mar 2005, p. 95.

5. Chahal SK, Environmental ethics, op. cit, p. 108.

6. Ibid p. 107.

7. Amartya Sen, The Idea of Justice, Allen Lane, Penguine, 2009, p. 251.

8. Ibid p. 249.

CHAPTER14

Democratic Values and Environmentalism

Economy and environment are standing in opposition to each other. The developing political economy of the globe has impressed upon serious repercussion on environmental system. Environmental hazards and threats have posed seriously alarming question of viability of contemporary materialistic economy progress which is disrupting the life sustaining resources deeply. Can material progress be at the cost of life. It is a burning question before bioeconomists. Four big[1] environmental fears which have come on surface are

1. Natural resources are running out.

2. The population is ever growing, leaving less and less to eat.

3. Species are becoming extinct in vast numbers; forests are disappearing and fish stock are collapsing.

4. The planet's air and water are becoming ever more polluted.

The system of governance, agendas, issues, aspirations and social web will remain only when the life tends to exist, if this is enkangered the everything goes a loss. The contemporary mandate of democratic governments and belief in democratic values gets lost if life will not remain. Human activity is thus defiling the earth, and humanity may end up killing itself in the process.[2]

Environmental problems and widening quanta of pollution and loving attitude towards consumerism have affected the democratic values and practices. The basic values of democracy – liberty, equality and rights can not be cherished in uncontrollable way by any nation, or any individual entrepreneur because if this be allowed for every owe, the chain of production and booming economy, the natural resources and life sustaining factors will get diminished very soon. So there is now the voices of putting ban on democratic guarantee of economic prosperity. No politics or political movement is allowed in this sphere now." It has become necessary for peace loving and ecological minded people to articulate not just a new politics, but a new ethics and a new earth centred moral sensibility that can reawaken the life affirming impulses our society seeks to submerge.[3]

And by virtue of fear of sustenance the life and resource scarcity for population, the need for severe political control, economic restraint and enforced uncharacteristic of liberal – democratic societies and capitalist economies has taken shape.[4] The spirit of democratic entrepreneurship and liberalism be shaped with the basic facts and essential constraints of environmental safety, is the voice of globe for cause of life itself now. No more interference be with the nature, if future generations are to be seen. For a democratic world order and sovereign states as independent player, up to the seventics of last century (twentieth century) there was as such severe constraint on using the natural resources, but after eighties, the growing momentum of saving environment caught hold and liberalism received the set back. Certainly for the green movements various factors as mentioned supra were responsible.

Environmentalism believes in solving the environmental problem with aid of technology and new methods and want to maintain the present method and values of production and consumption[5], and as it goes on a natural police state emerges. The reason is that for increasing economy an increased watch and monitoring of state is required, and this automatically implies the gradual emergence of bureaucratic police state[6]. This police state in turn proves a dictator for democratic values. Though uncontrollable democracy is never advocated in any era yet under some reasoned and rationale control of sovereign (state) it was much more wider than the recently emerging police state. The emergence of bureaucratic police state has the economic and political reasons as caustive factors. Growing and blooming economics have certainly environmental negative impacts.

Democratic states have always believed in going just and timely representation of governees in government, The decision process in carried out by taking the majority consenous in any form. The environmental issues have also been processed through these mechanism, but environmentalism has demanded the inclusion of environmental agenda as the prime agenda in forming any government; there, at the same time, a demand for openness of govt. (state policies) regarding environmental issues has existed. "There has all along among environmentalists been a very strong sense of the public's right to know and to be involved in the decision that affect their lives.[7] Thus environmentalism has supported the values of participatory coherence for a better eco-democratic society. "Innovative public participation in the environmental policy arena has taken other forms as well. These will include such diverse initiatives as: (1) community right to know legislation, (2) Internal responsibility systems regarding occupational health and safety, (3) the use of referenda in environmental matters, and (4) direct participatory interaction between environmental citizen organization and co-operations.[8]

There has been a change on ecologists and up to certain extent on environmentalists for reducing the opportunities of employment. The developing economy provides better opportunities for employment a basic tenant of democracy but a ban on the economic activities as ecologists suggest will cause unemployment. But this is only one aspect of coin, the other aspect which "Greens" advocate in rational use of local resources with global concerns. The local communities generate employment activities automatically. In fact the shift of burden of giving employment to local or natural repository is guided by the philosophy of human being the part of nature.

The democratic theory of state or in practical term, governance not only covers the polity and statecraft but it extends to the nature and its inherent values. For the cause of good life and responsible well being with social capital and sustainable livelehood[9] for future, the commitment to certain values is essential as under[10]

1. Reducing consumption, waste and the general exploitation of non-human nature.

2. Allowing ecological systems to work within their carrying capacity and biological diversity to be preserved and ensuring resources are not used at rates faster then they naturally reproduce themselves.

3. Fostering a way of life that incorporates patterns of consumption and production which contributes better to human flourishing.

Eco-democratic values also get nourishment from to be rather than to have[11]. It is an eastern spiritual dictate. This attitude eradicates the factor of ruthless need fulfillment and promotes contemplation.

Adding this value, "Greens" also believe in fantastic democratic value of 'thick globally and act locally' Green political action, democratic participation in local issues and global concerns, decentralization, self-sutaining and self-regulating local bio-regional economic and administrative units.[12]

The protection of environment will mandate the economics and finally the globalisation, is a truth, and it is being mandated by "Greens". The relationship among resource limitation, environmentalism and democratic values for a common good is unfailing consensus of present generation – directly or indirectly. No one wants to die and to not have their children of future.

The question of ecologism as an ideology and its values of fostering deep "Greenism" or a bit shallow adjustment with the former and or environmentalism – the managerial technique to live with present modes and values of consumption and production, find ultimately the fine tuning of democratic values, but these democratic values are "Green" as explained supra.

The eco-democratic values have also witnessed mass mobilization. Democratic mobilization is essential to the achievement of such policies which point to sustainability, decentralisaion of economics and employment opportunities and enhancement of opportunities for democratic participation[13]. Environmentalism must be on continuance for inculcating the democratic values for a better environmental society, however it will not be a purely ecological value based society – it is another fact of whole story.

And last but not least the genetic value of eco-democracy is our moral responsibilities to future generation. How resources should be used is more important than to consider who should use them. Ethics in allocation, in other words is not a consequence of ethics in distribution. An environmental ethics cannot be derived entirely from a theory of justice[14] the backbone of democratic society.

REFERENCES

1. Bjorn Lomborg, The Truth about the environment, in John S Dryzeck and David Schlosberg, Debating the earth Oxford, 2005, p. 74.

2. Ibid p. 74.

3. Agarwal Sk, Green Management, APH Publishing Cooperation, New Delhi, 2005, p. 57.

4. Robert Pachlke, Democracy and environmentalism, in John S Dryzek and David Schlosberg (ed), op. cit p. 164.

5. Andrew Dowson, Environmentalism & Ecologism, in Alan Finlayson (ed), Edinburg University Press, 2003, p. 385.

6. Passmore, man's responsibility for nature, Duckworth London, 1974, pp. 193-194.

7. Robert Pachlke, op. cit p. 164.

8. Op. cit p. 171.

9. Robert Chambers, ideas for development, Earthscan, London, 2005, pp. 200-201.

10. Tim Hayward in Allen Finlayson contemporary political thoughts, Edinburg University, 2003, p. 357.

11. Op. cit p. 358.

12. Robert Goodin, Green Political Theory, Cambridge Polity Press, 1992, pp. 13-16 and 113-68.

13. Robert Paehlke, op. cit p. 175.

14. Mark Sagott, The allocation and distribution of resources, in John S Dryzeck and David Schlosberg, Ibid p. 156.

CHAPTER-15

Conservation in Green Political Theory

Green Political Theory is an elaboration of ecoligal footprints and ecological footprints are major issues of concern when we deal with the crisis of consumption and conservation. The equation of consumption and conservation seems linear but infact it is neither linear nor circular. Infact it is beyond consumption and conservation. The crux of the crisis rests with ethos and life living socio-political philosophy and ethics. If the stand is on ecocentric philosophy the picture of conservation is peculiar allowing let the nature run on its own way (whether human interferance and actions are natural is a question of debate but for the cause of survival and beyond certain limits they canbe put unnatural), if the stand is on anthropocentric philosophy, the case of environmentalism comes foreward and if the stand is on biocentric philosophy the focus comes between ecocentric and anthropocentric philosophies.

The ecocentric determinism deserves case and consideration in itself. The case and conservation is neither liberal nor utilitarian. It is by virtue of its stand. It is preservation of entire planet with all its creation for all time. The anthropocentric view has liberal and utilitarian view. For cause of human and in interest of humans the conservation of natural resources and wild is essential. And the biocentric view cases for conservation of all forms of life, because they have an inherent right to exist and live[1].

The issue of consevation is a metastasis of consumption, because the above referred views also define the behaviour and pattern of consumption. Biota, Biodiversity and corresponding ecological niche decide the ecological foot prints. The crisis of conservation has an inherent value and its major sheet anchor is role of human as biota.

The preservation of biodiversity is not the sole concern of conservation, at the same time the major problems of Pollution, deplection of non-renewable resources and degradation of renewable resources[2] are there "The Contemplation of the word's disappearing supplies of minerals, forests and other exhaustible. assets has led to demands for their regulation and exploitation[3] and not only this, the conservation of biota is also a demand against their extinction.

In nut shell the compartmentalization of conservation issue can be sketched in – (1) Prevention of biotic resource against their extinction (2) prevention of non-biotic resources against their exhaustion and depletion. Prevention is a negative rather neutral concept. It defines the regulatory action. When the same consurvation is viewed from point of view of preservation, it is ethically a positive concept. It requires extra efforts beyond prevention. In the same continuity the consumption behaviour will have to be remodeled accordingly. And the question becomes more interesting when the light comes from different angles of eco centrism, anthropocentrism and biocentric anthropocentrism.

Ecocentric aspect allows the preventation as well as preservation both and not only this, it also allows culmination of nature as an independent entity. It is crystal clear that social and political framework and economic activities will have to be redesigned accordingly. The human centredness is not a pivotal factor here. The eccentricity of conservation is towards the nature as a goal and inherent value. The anthropocentric aspect keeps the human centredness in focus and then defines prevention and preservation. A selective approach of conservation is manifested here. For the utilitarian element the conservation comes here and prior to that consumption according to human need is present before the issue. Biocentric aspects is partial. It cases for biotic category of conservation. The unloving resources are not here and at the same time it is half ecocentric and half anthropocentric.

One more ontological issue of green political theory is definition of "nature", Nature is segmental and at the same time by virtue of philosophical footing it encompasses the whole including human action. "Nature acts outside our plan. We must be prepared to allow nature to act. Our conservation must recognize that it does so and not seek to conrral it within the bounds of either our predictive conservation science ....[4]". All is embodied by nature itself be it anthropocentrism, ecocentricism or biocentrism, this is one aspect to view the nature. It runs into "auto mode". "Let it be" is there as it is there. But if nature is viewed apart from human actions and focused towards virginity of natural biotic, and abiotic resources, the eccentricity of conservation is spread as depicted above.

It is also perceived that when we talk about nature. It is understood wild, virgin wild, but when we perceive environment it is "nature manned with" and the politics associated with it[5]. It can also be understood, when we echo ecology it is "wild nature", and the perception "what is happening in all" is nature. So a subtle difference is there, and accordingly the perception of conservation will be arrayed. It is important to note that in green political theory the concept of conservation is deeply attached with socio-political impact of "nature" and its expression.

The political aim of citizenship is good life, as per Aristotle, but it is only possible when life remains entact. The "Gaia" theory which was propounded by James E Lovelock in 1976[6] exhabits stability of certain environmental aspects and state. But the economic and political interests are ruining the equitable sub stales to remaining depletion of environmental aspects. In this tune "Gaia" suggests that a precautionary approach is important on large geographical scale, as well as at the level of loss of individual species. It highlights the interdependency of species. Ultimately this may prove one of the most compelling reasons for conservation. [7]

Gaia hypothesis thus not only hints for sustainable theory of conservation but also the justice of conservation, preservation and promotion. It also sweeps the general liberity of life and existence and thus gender issues in whole and at the same the right to life, if we concentrate on biotic aspect of conservation. Because Gaia suggests existence of biological diversity thus it stands in favour of rights of human as well as non-human living world to lead a blissful life.

Thus why conserve has different stands. The utilitarian and the intrinsic value aspects are there besides being the sustainable philosophy of sustenance of life. The sustainable philosophy of life can be eco-centric or a mixed state with other two aspects anthropocentric and bio-centric. But a way of life which emphasis inter-connectedness, decentralization, non-harmfulness to other life and respect for the rights of other organissm, low consumption and feeling of unity with rest universe. This is stand of deep ecology[8] but at the same time when this sense mixes with flourishment of every creature and aesthetical beauty culmination of every constituent, it reflects the spirit of Sarvodaya. "The etymological meaning of Sarvodaya is rising of all. This all includes all living beings.[9] The Sarvodaya aspect of conservation is one step ahead. That brings not only the attachment with rising of all but also detachment to self-lust and renunciation of desires. The conservation, thus has a spiritual look in this background. The platonic detachment[10] (Gita's anasakti) with the surroundings but having the feeling of greatest good of all ultimately writes the spiritual theory of conservation.

The theory of conservation in thus contemplated in terms of economic activities but reshaped with philosophy of socio-political relationship. Green political theory, infact, is a wonderful multidisciplinary gamut of conventional political philosophy – the very state where various subjects contribute.

The aspect of conservation is not empirical but normative. Leaving the utilitarian and liberal concepts of political philosophy and keeping aside the capitalistic module of thinking "Permanent support for conservation will depend on the success of an appeal" to a deeper feeling for nature as something valued for its intrinsic interest and inspirational significance. People need to be persuaded to "respect nature for its own sake" [11]

Green Political Theory also demands that the sense of conservation calls for the close contact of people with nature, and mass movements to ignite the light of intrinsic value of that nature which is something other than human deteriorative actions; maybe the economic developments are hamphered.

Cherishing the essense of conservation[10] the green citizenship also commands the observance to same.... "Citizenship not only concerns right and entitlements, but also duties, obligations and responsibilities. [12]

It is quite interesting to note that the ecological citizenship is not citizenship to any particular territory but in the ecological chase the relevant space of citizenship obligation in best expressed via the earthy notion of the "ecological footprints". [13] The concern of conservation is thus tracking the "ecological footprints" with a share and care of present and future generations of biotic world as a whole and non-living world in its full conformity. The sustainable philosophy of life which continues for ever beyond the realms of polity and economics just for the sake of life is the crux of conservation and its issues.

REFERENCES

1. R. Rajgopalan, Environmental studies, Oxford University Press, 2005, pp. 327-328.

2. Ulaganathan Sankar, Environmental Economics Oxford University Press, 2001, p. 398.

3. Ibid, p. 83.

4. WM Adams, future nature, earthscan London 2003, p. 170.

5. Wolfgang sach the development dictionary, orient longman, 1997, p. 46.

6. 1

7. Clive Hamber, conservation, Cambridge University Press, 2004, p. 37.

8. Op. cit p. 24.

9. Anil Dutta Mishra, Fundamentals of Gandhism, Mittal Publishing Co. New Delhi, 1995, p. 5.

10. BB Pandey, Gandhi, Sarvodaya and organization, Chugh Publication, Allahabad, 1988, pp. 20-21.

11. WH Adams, Ibid, p. 109.

12. Van Steenbergen, B. Towards a global ecological citizen in bart van steenbergen (ed), The condition of citizenship, London, Sage, 1994, p. 1466.

13. Andrew Dobson Ecological citizenship (ed) in John S Dryzen, David Schlosberg, Ibid p. 598.

CHAPTER-16

Ancient Indian Experience to Green Political Theory

Indian society has shown a deep concern regarding the ecological issues since its dawn. The nomenclature may be different but the essence of Green Political Theory is very much there in Indian classical texts.

"The Rigveda, one of the foundational texts of Indian view of life presents a world view where is man is a part of the cosmos."[1] Rigvedic poet assumes man to be a child of mother earth. The respectful homage to earth and sacredness of earth as mother is noting eco-philosophy of Rigveda. If the divinity is attached with natural objects, subjects and various phenomena, the behaviour to them gets changed. It can be assumed that in such state of efforts, the ruthless and brutal exploitation of earth and its natural resources and inferior sentiments to other non-human living species/beings can not take place.

The environmental issues are multidimensional issues and require a close watch: but in Indian context the culture of India has casted a deep influence on environmental awareness. "Indian culture is based on ethics nature and its segments treated as a holy symbols of God and Goddess"[2]

The prithvisukta of Atharvaveda reinforces the idea of man being the son of earth.[3] And the "Rigveda frequently involves plants, along with rivers, mountains, heaven and earth as divercities"[4] The Rigveda also coins the concept of "Rita" which is eternal truth, and this truth eternal world order; In this world order every constituent of cosmos has to maintain the discipline and remain at disciplined space. This world order, which is a divine law, binds every one in one thread of oneness.

The Indian classicism is overwhelmed with equal status for all creatures. Man has not been given the superiority over other creatures. Atharvaveda[5] echoes the earth is for all the creatures, and not only for human beings alone. God is one and it is manifested in different deities. The creation and creatures are partial part of God. The same spirit is echoed by Rigveda[6], Yajurveda[7] and other later texts. The same spirit of God manifestation has come in Mahabharta[8] and shrimad Bhagwatgeeta. [9]

Indian experience of the same soul in every creature is vividly manifested in not only assuming oneness of whole of the world and its inter connectedness but also putting the deep regards to non-human world." Like the trees and forests, the river system which have been the arteries of India's livelihood have also been an object of workship from time immemorial.[10]"

The pious sects and sacred rituals flourished in natural ecosystem in India the forests, trees, natural powers of stars and natural phenomena were all dwellings of these activities. In fact the whole life of ancient India was moving with nature and love to nature. The Indian classical texts even dealing with whole philosophy of livelihood and spiritualism were turned as Aaranyak.

In Aaryan civilization the animal sacrifice was the sacred ritual but later on when the civilization intermingled with post vedic period thought, the killing of animals were prohibited.

Different animals were said to be produced by almighty in different texts such as Mundakopanished," Shri Madbhagwat puran, Manusmrati[13]. and Arthashastra[14]. Animals and birds received respect and even Hindu Gods were using them as their vehicles. The animals not only claimed their existence in Indian classical society but also claimed the sacred status. In the same tune Vishnu Purana says the love going to all animals. It "Continues" Who does not harm or destroy other non-speaking creatures or animals god loved them" [15] Thus it is clear that the claims and rights of animals as are advocated by modern ecologists find the place in ancient Indian texts. The feeling of emotions were not found only in animals but also in trees. Mahabharata observes that trees have emotions and the man who cut them causes them sorry. Various examples of tree worship, tree-brotherhood and tree-companionship are found in ancient Indian texts. The right of trees which have still not been advocated vehemently by putting them as living entities were championed at that time. Modern Green Political Theory Stands for rights of unserved animals and tries to keep them away from pain, but the essence of this spirit was mandatory for kings to be implemented. Various examples of such nature can be cited in Ramayan and Mahabharata. There are elaborate mechanism in Kautilya's state to present cruelty to animals[16].

Green political theory also encompasses the afforestation and avoidance of pollution. The afforestation and avoidance of pollution. The same has been taken case in ancient India. Matsya Purana[17] says those who plant the trees go to "Moksha" (complete salvation). Other puranas also echo in same tune. The tree workship inheritage from Vedic period to later ages continued without interruption.

Green political theory underlines the issue of balanced use of natural resources in shape of environmentalism and planning. Indian intelligentia had observed this view quite earlier, and we find the accounts of the same spirit in ancient works of Vastuvidhya. "It seems that ancient works on Vastushastra were aware of the ecological problems caused by the pressure on land due to unbridled construction work.[18]"

Indian mind was so over whelmed with the sensitivity of environmental issues and pollution it repeatedly cared cleanliness of mind because if mind is clean and full of pure and devine thoughts the pollution can be controlled. The texts of "Charak

Samhita" and "Sushrut Samhita" cared this aspect very religiously. Mahabharata[19] also underlines the same idea.

In the same continuity various citations can be quoted but it is revealed that following dictates are very much there in Indian experience.

1. Earth is sacred and devine.

2. Nature is bestower of all blessings to human and non-human world.

3. Every constituent of existence should be looked with reverence and equal status.

4. Rivers, Trees, Animals, Birds, Natural resources, non-human beings have been given their own appeal of being.

5. The moral virtues of harmony have been continuously inculcated in minds of inhabitants so that automatically environmental ethics may take place.

6. The life philosophy of Indian has been shaped in such a way where simplicity, transparency, renunciation and spiritual development have been kept at top. The philosophy to life living is leading to harmony and cooperation and not to violence, exploitation and sinful procurement.

The above accounts show that the essence of ecologism has appeared in Indian thoughts. The "Spiritual Ecologism" can be a better name of Indian Green political theory in its classical sense.

REFERENCES

1. Ranjan Chakraborti (ed), Does Environmental History matters, Reader's service Kolkata, 2006, pp. 89-90.

2. M.V. Joshi, Environmental Economics, Atlantic Publishers and Distributor, New Delhi, 2001, p. 205.

3. AN Yellappereddy, Role of Sanskrit in preservation of wild life Biodiversity and environment, Souvenir, LBRS Vidyapeeth New Delhi- ch. 184.

4. Rajan Chakroborti, op. cit p. 90.

5. Atharvaveda, 12.1.15

6. Rigveda 1.164.46.

7. Yajurveda, 32.10.

8. Mahabharata 182.14-19.

9. Bhagwat Geeta 7.6.

10. Ranjan Chakraborti

11. Mundakopanishad 2.2.7.

12. Shrimad Bhagwat Puran 1.5/7.14.9.

13. Manusmrati 5.45.

14. Rang Rajan (ed), Arthashartra, Delhi, 9.-96.

15. M.V. Joshi op. cit 215.

16. Ranjan Chakraborti op. cit p. 93.

17. Matsya Purana 59:159.

18. Ranjan Chakraborti, op. cit pp. 95-96

19. Mahabharta 16.8.11.

CHAPTER-17

Gandhian Ecological Society : A Paradigm

Gandhi has not said directly on ecological political theory but he was very much concerned with the issues of the same. Gandhian philosophy of truth, non-violence plain living and high thinking, the ideal of socially ethical and altruistic life frames the complete saga of his ideal society and ultimately his understanding defines the characteristics of ecological society. Before defining ecological society we will look at civil society. The civil society is the sphere of civil autonomy.[1] The civil society is an idea of balance between individualism and collective life and evolving the true spirit of democracy in managing the affairs of Civility. It has been felt that in name of democracy the particular class (political class) rules and the civilians are left. Civil society allows full participation and decision making in affairs and so it also negates the state authority. "Civility as a future of civil society considers others as a fellow citizens of equal dignity in their rights and obligations as members of civil society[2]".

Haberman forms the civil society as a site where common understanding about public goals and exercise comes through the forgive of individuals opinions.[3] Civil society thus fosters democracy by limiting the state. The normative view of civil society underlines the existence of democratic interaction, trust, tolerance, cooperation and equality.[4]

Thus we see that apart form different interpretion of civil society the basic elements of civil society are – democracy in true sense and participatory responsibilities evolution of democratic values of mutual respect equality, tolerance and trust, cooperative economy for common good, self management and checking the state autonomy.

When we concern the ecology we find that the world has been coined by taking derivations from the Greek "oikos" meaning household or habitat[5]. Habitation has meaning of a cooperative living the living with organic and non-organic and human and non-human world. Later on when in late twentieth century ecologism grew inform of ideology, it stressed the central importance of ecology, it highlighted the re-thinking of the relationship between human being and the natural world[6]. It is in fact totality of relationships among biotic and abiotic constituents[7].

Before the characteristics of ecological society are referred the essential elements of ecologism should be mentioned and that are ecology, holism, sustainability, environmental ethics, and self-actualization[8]. The ecological society should contain in themselves these elements. The essentials of civil society have commons with ecological society and it is apparently clear that normative aspect and values of civil society, evolving and participating responsible democracy and ethical composition of individual tending to common good are but essential aspects of ecological society. It has also been opined that over emphasis on values in civil society may distort the democratic process[9] but the factor of altruism subsides the whole matter : The factor of altruism and tolerance to others has a major say in ecologism.

Ecologism has a sizeable concern with ecocentrilism with much hue and cry[11]. Deep ecologists dismiss shallow ecologism as a thinly cancelled form of anthropocentrism, arguing that its central objective is to maintain the health and prosperity of people who lived developed countries. The 'deep' perspective completely rejects any lingering belief that the human species is in some way superior to or more important than, any other species or indeed nature itself[11] but the pliant anthropocentrism which, on the practical ground echoes codependency of nature and human, though filtered through human interest[11] frames the ecological structure changed with minimum intervention of human.

Ecologism has viewed nature as mother and seen the permance of survival in eastern spiritual ideas. Sustainability is a major concern for ecological society and "Buddhist economics based upon the principle of 'right livelihood' and stands in stark contrast to conventional economic theories, which assume that individuals are nothing more than utility maximizers. [12]

Ecologism has also sees the moral stance of "biocentric equality" Arne Naess[13] has expressed this an equal right to "live and bloom" .

The ecological society has all these traits. Besides this the agenda of ecological movement that is decentralisation of activities but thinking globally have the vital practical role.The decent ralised self-sustaining, self-sufficient and regulating and self governing society which acts locally but thinks globally[17].

Gandhian notion of ecological society (which is very much manifesting the civil society) is "willing submission to social restraint for the sake of the well being of the whole society enriches both the individualism that is justifiable that which is built up in the service of others" [15] Gandhi advocates for altruism but his "swaraj" is a sacred world; means self role and self restraint[16] Gandhi in his society also talks about the rights of disabled and animals. His non-violence theory supports the course of every constituent of nature.

Gandhi's civilian in self-rule is not selfish. Gandhi civilian's value depend upon his way of life not his status. One's way of life is not to be judged by one's. Visible outwards action, but by one's inner learnings.... To sum up he alone is happy and he alone is wealthy who is sincere in himself, bears no malaise, exploits no one and always acts with a pure mind. Thus Gandhi's ecological society will be inhabitated be pure minded altruistic civilians.

Gandhi elaborates " let there be no mistake as to what purna swaraj means ... It is full economic freedom for the toithing mission. It is no unholy alliance with any interest for their exploitation"[18]. Gandhi understands quite in advance before ecologism movement that exploitation will never allow the self regulation and self-management. The exploitation of natural resources, as is being seen today, can never bring ecological society, because it obstructs the basic fabric of self regulating local communities and brings ecological devastations.

The ecological society advocates a natural life which respects nature in its virginity and if human interests are there they are well reconciled just for plain living. Gandhi opines "For me the only training in swaraj we need is the ability to defend ourselves against the whole world and to live our natural life in perfect freedom, even though it may be full of defects. Good movement is no substitute for self govt. [19]"

Gandhi believes in self sufficient village. He sees that this village community should be self sufficient in basic necessities. My idea of self-sufficiency is that village must be self-sufficient in regard to food cloth and other basic necessities... Self-sufficiency does not mean narrowness. To be self-sufficient is not to be altogether self contained[20].

Gandhi's self-sufficiency and self-regulation goes up to the extent when he says "Every village has to be self-sustained and capable of managing its affairs even to the extent of defending itself against the whole world" [21]

Gandhi's village ecolomy is thus simple, satisfying and suiting to bare needs only. Gandhi's philosophy shows the devastation of consumption economy and the greed element. He says "there is enough in the world to meet the needs of every one but there is not enough to meet the greed of everyone"... Therefore he did not advocate macro economic policy. All villages should become self-sufficient. It must work from the bottom upwards and not the top down[22] Thus Gandhi in the very stream opines self-sufficient decent ralisation – which is the key element of ecological society. The local concerns of village community will never go for the massive exploitation which is causing the degradation of nature and its virginity.

Gandhi believes in "sarvodaya" upliftment of all. On the basis of truth, non violence, love and non-exploitation, the 'society must strive for the economic social spiritual and physical well being of all, not only of just the majority. He favoured a holistic approach to well being, and a total approach to community.

Gandhi identified the ill effects of all these events and took the stand "He believed that there should be an equitable distribution of resources and that communities must become self-sustaining through reliance on local products instead of large-scale imports from outside. Gandhi was opposed to large-scale industrialization and favoured small scale industries that promotes local self-sufficiency, which he caller swadeshi[24]"

Gandhi's theory of trusteeship of resources paves the stranger foundation for ecological society, because the element of exploitation, conspicuous consumption, materialistic, slow off etc. wash away. The ethics of sustainable development of ecological society lies in Gandhian thought when he says "Rural economy as I have conceived it, eschews exploitation altogether, and exploitation is the essence of violence[25]" and "Under the Gandhian economic order the charter of production will be determined by social necessity, and not by personal whim or greed[26]. The present generations desires and future generations needs are well woven in Gandhian socio-economic philosophy of village self sufficiency non-violence and trusteship theory.

Gandhi is so impressed with self sufficient, self-governed village panchayats (the local community) that he does not adopt state authority. For him the state is replica of violence. The local community understands its needs well thus it automatically preserves the nature. The state on horse of power and violence may distort the nature. Ecologiacal society is eco anarchistic. "Anarchist believes in stateless society, in which harmony develops out of mutual respect and social solidarity amongst human beings. ... Ecologists also believe that balance or harmony, spontaneously develop within nature, in the form of ecosystem and that these, like anarchist Communities, requires no external authority or control

....Anarchists have also advocated the construction of decentralized societies organized as a collection of communes or villages. Life in such communities would be lived close to nature, each community attempting to achieve a high degree of self-sufficiency[27]. Gandhi believes in such village (Society). "Legislation imposed by people upon themselves is non-violence to the extent it is possible is society. A society organized and run on the basis of complete non-violence would be the purest anarchy"[28].

The need of ecological society is intense and highly desired because the natural resource depletion, ecological devastation and pollution are on increase and this phenomena is questioning the survival of whole planet. When the "specters of global warming, lack of water through deforestation and continued depletion of natural resources and diversity on earth are some of the results of 'Unsustainable development' and economic growth. Should we not follow the path of Gandhi?".... We are slowly realizing that reducing poverty or moving towards sustainable development is not just an economic or a technical problem or one of acquiring greater financial inputs. All these are important but achieving these goals also needs an inner awakening an inner transformation of man – path already paved by Gandhi[20]

The practical concern here is that when the contemporary world has advanced too far in direction of industrialization. Globalisation and consumption economy and segregated community is not possible in these situation, then Gandhian self managing rural economy and thus formed ecological society will do, in a question. The aswer may be searched in Gandhian philosophy itself, when Gandhi opines the ideal of creating an unlimited number of wants and satisfying then seems to be a delusion and .... This philosophy can only bring ecological society in reality.

The ecological society finds its best in Gandhian philosophy and it is assured to be names as Gandhian ecological society, though Gandhi never used such term in his words

REFERENCES

1. Boker Gideon, and society and democratic theory, alternative voices, Routledge London, 2002 p. 10.

2. Sumeet K. Pathak, Gandhi and Civil Society, The Indian Journal & Political Science Meerut Vol., LXIX no. 2 Ap.-June2008, p. 271

3. Carolun M Elloitt, Civil Society and Democracy, Oxford University Press, 2003, p. 7.

4. Ibid p. 18.

5. Andrew Heywood, Political ideologies, Palgrave Marmeallan, New York, 2003, p. 266.

6. Ibid pp. 266-267.

7. Trim Hayword, Ecologism and Environmentalism in Alan Finlayson (ed), contemporary Political Thought Edinburg University Press, Edinburgh, 2003, p. 355.

8. Andrew Heywood op. cit, p. 269.

9. Juan Linz and Alfred Stepan, Problems of Democratic Transtion and Consolidation (The Johns Hopking University Press Baltimore, 1996) p. 10.

10. Andrew Heywood op. cit p.-271.

11. Andrew Vincent Green Political Theory, in Richard Bellamy and Andrew Mason, Manchester university press, Ny 2003, p. 184.

12. Schumacher FE, small is beautiful, Blond and Briggs, London, 1973.

13. Arne Naess, Community and life style, outlive of an ecosophy, Cambridge University Press, Cambridge, 1989.

14. Robert Goodin, Principles of Green Politics, in Alan Finlayson, op. cit., pp. 398-404.

15. Biswas S.C. (ed) Gandhi, Theory & Practice, Social Impact and contemporary Relevance, Indian Institute of Advanced studies, shunle; Kp Boychi & Co Kolkata, 1969, p. 322.

16. MK Gandhi, Young India, 19/03/31, p. 38.

17. Shriman Narayan, The selected works of Mahatma Gandhi IV, Navjeevan Ahmedbad, p. 34.

18. MK Gandhi young India 16/4/31, p. 77.

19. MK Gandhi Young India, 22/03/20, p. 1.

20. MK Gandhi Khadi why and how, Navjeevan Publishing, Ahmadabad 1959, p. 166.

21. MK Gandhi, Harijan 28/03/46, p. 236.

22. Rana P.B. Singh Mohandas Gandhi, M David Simon (ed) fifty key thinkers on development Routledge Lond New York, 2006, pp. 108-109.

23. Ibid p. 110.

24. Ibid p. 110.

25. MK Gandhi Harijan 4/11/39, p. 331.

26. MK Gandhi Harijan 25/10/52 p. 301.

27. Andrew Heywood, op. cit p. 287.

28. MK Gandhi, harijan 21/7/40, p. 211.

29. Ram P.B. Singh op. cit pp. 110-11.

30. MK Gandhi, Harijan, 29/8/36 p. 226.

CHAPTER-18

Gandhian Consciousness of Green Political Theory

Mahatma Gandhi never claimed about himself as a political philosopher. He was a man of action who believed in transforming the world through implementing his ideas. Any Gandhianism exists was doubtful in Gandhi's eye, but Gandhianism do exist. Gandhi's idea of truth and non-violence have stirred the whole political thought Gandhi says ' We have to make truth and non violence not matters for mere individual practice, but for practice by groups and communities and nations" [1]

For Gandhi truth is God and when one wants to find the God, love is the means, love is non-violence[2]. Gandhi was an ardent ethical person and his ideas of socio-politico economical sphere root from the sphere of spiritialism and ethics. Gandhi's non-violence not only limits itself with human world but it comprises of the whole creation. "Ahimsa includes the whole creation and not only human[3]".. Gandhi's this idea gives a unfailing support to those Green political thinkers who can never been truthless and virlent exploitation of ecology and devastation of non-human world. Gandhi's "Ahimsa" is not only limited among human beings "in its positive form, Ahinsa means the largest love, the greatest charity" [4]

The Green political theory not only rests with ecocentric or anthropocentric centers but it subtlěy underlines the nature of man. The violence of human nature and that too out of greed manifests violence and devastation in non-human world. "We are cutting trees faster than they can regerate overgrazig rangelands and converting them into deserts, over pumping aquifers; and draining river dry .... We are taking fish from the ocean faster than they can reproduce .... We are releasing carbon dioxide into the atmosphere faster than nature can absorb it .... Habitual destruction and climate change are destroying planet and animal species far faster than now. Species can evolve[5]. This whole scenario puts the violent nature of man which has now become a big threat to existence of man himself. Gandhi at this juncture hits the target when he opines "Man's nature then is not "Himsa", but "Ahimsa", for he can speak from experience his inner most conviction that he is not the body but atman. [6]".... He keeps on saying that "Conquest of one's passions, therefore is not superhuman, but humane and observance of Ahinsa is heroism of the highest type" [7] Quite eloquently and keeping himself in to with ecological relevance Gandhi further echos "When non violence is accepted as the law of life, it must pervade the whole being and not be applied to isolated acts[8].

Violence and embedded with it the brute and greedy nature of man can disrupt the ecology and environment and thus the ecology and environment and thus the survival of human an other beings, in continuity to this dictum one of the deep attraction of Green political theory is its claim to be focused on the very survival of the whole natural eco-system of the planet. In consequence, it also addresses the condition for our biological continuance as a species .... Green political theory conceives of itself as "green", 'environmental' or 'ecological', because of its key focus on nature .... Nature qua green, is the key theme[9]"

Gandhi is much advance in balancing the nature, when he clings to the basic unit of human hood that is man's nature to be non violent. Gandhi says truth (the replica of God) in center of creation[10]. Gandhi supports ecologism and "green" when he opines "I find in the practice of tree worship a thing instinct with a deep pathos and poetic beauty. It symbolises true reverence for the entire vegetable kingdom[11]. Gandhi continues "I tries to see him and his mercics in all these creation[12].

The concept of nature itself is not clear in Green Political Theory, at one hand it assumes the humans as knots in great ecosphere net on the other the human and its actions integral part of great nature. The first assumption underlines the ecocentric value that is ecocentricism and human existence as cooperating knot to great ecological sphere the other assumption is anthropocentric and tries to manage the increasing desires of man and present mode of production and consumption with the environment, means the ecology be adjusted according to man's action. This is pliant anthropocentrism stresses co-depending with nature, although still filtered through human interest[13]"

It is apparently clear that Green political theory has moved in special dimension where the "nature" is co-read with socio-politico economical and other allied concepts. Though Gandhi does not say specifically about Green Political theory but Gandhian thought takes every care of the same. Gandhi understands that codependency with nature is unavoidable. The modern concept of pliant anthropocentrism or ecologism get the best balance in Gandhian thought. "Environmentalism argues for a managerial approach to environmental problems, secure in the belief that they can be solved without fundamental changes in present values or pattern of production and consumption and ecologism holds that a sustainable and fulfilling existence presupposes radical change in our relationship with the non-human natural world, and in our mode of social and political life"[14]. Gandhi understood both the approaches quite earlier and after examination he opined"... the ideal of creating an unlimited number of wants and satisfying them seems to be delusion and a square"[15] Gandhi believes in plain living and high thinking "The less you posses, the less you want, the better you are[16]. He goes further and says "High thinking is inconsistent with complicated material life based on high speed imposed on use by mammon worship. All the graces of life are possible only when we learn the art of living nobly"[17]. One step going ahead and putting the economics in this framework but also keeping the man's real nature of possession Gandhi gives the concept of trusteeship. This theory dictates that by virtue of man's entrepreneurship earn as per social need not out of greed and he or she is allowed for that but he or she will use just that necessary proceed which is really necessary for living a plain, and ethical life and the remaining will be the trust part of other." The transformation of private property into public property has been achieved by the application of doctrine of trusteeship"[18]. It is a unique experiment. In favour of nature (which is a truth and truth in God) Gandhi opines everything belonged to God and was from God. Therefore it was for His people as a whole not for a particular individual. When an individual had more than his proportionate portion he became trustee of that portion for God's people" [19]

Green political theory indirectly underlines the element of non-possession. Green is a concern for this theory and Gandhi had identified this element quite earlier. The element of greed and possessiveness had harmed ecology. :The ecologism represents a new style of politics. It starts not from a conception of humanity of human needs, but from a vision of nature as a network of precious but fragile relationship between living species including the human species and the natural environment[20]. Gandhian idea of truth, non-violence, ethical life, plane living and high thinking and production as per social need and not out of greed and finally trusteeship theory make the whole gamut where Gandhi can stand with any devasting anthropocentric concept to basic structure of green political complex.

Green political theory talks about the ill effects of indistrialisation when Lester brown[21]. Counts that consumption of earth's natural resources is giving birth to bubble economy. The hazards to natural resources, change in climate, ozone layer depletion, acid rain, expansion of deserts, scarcity of water, lowering of water table, and increasing population are all appearing before us. This is result of hyper industrialization, and anybody will agree that these are evils of industrialization. Gandhi opines "Industrialism is, I am afraid, going to be a curse for mankind. [22] Gandhi is against machinery which is not subservant to interest of all 23 and destructive 24. Gandhi opted for that machine which is helpful for a self sustained village economy and which saves labour or under pains.

Green political theory also talks about sustainable development but what in the real definition of the same is still a question of debate, however, sustainable development is "development that meets the needs of the present without compromising the ability to future generations to meet their own needs. It contains within two key components : the concept of 'nests', in particular the essential needs of the world's poor, to which overriding priority should be given; and the idea of limitation imposed by the state of technology and social organization in the environment's ability to meet present and future meals" [25]. The inter and intra-generational justice glittering in the spirit of sustainable development shows that horizon of development and progress which should be sustaining the natural environment in co authorship of human actions. Many environmentalists have decried its 'anthropocentrism', deploring the fact that it unasamedty takes the promotion of human welfare as its central value. They argue that rather than warrying about sustainable development the should be more concerned with sustaining the natural environment[26] Gandhi's idea that "Man should live in cooperation and look for the common good"[27] Gandhi's common good is extended to whole of the creation extending to future generation. Gandhi opines" I do not believe in the "greatest good of the greatest number", nor can I agree that might is right. For human beings the object in view should be the good of all, with the weak being served first"[28] and "Please ponder over this form the point of view altruism, not of selfishness" [29] Gandhian view in the way takes care of the whole beings whether present or coming in future.

Green political theory manifest a spiritual dimension of human consciousness" It is founded on a recognition of the "oneness" of creation and a subsequent reverence for one's own life, the life of other and the earth itself" [30]. The common good to all is only possible when "oneness" exists. Spiritualism does not tolerate and that is advocated by Gandhi, and the violence on non-human world is also not permissible in Green political theory.

The central point of Green political theory is

1. Identifying the normative concepts of political philosophy with the third dimension of truth nature and its existence and inter and intra mechanism.

2. Identifying the man's nature.

3. The appeal of 'nature' and 'non-human livings' to human on holistic platform.

4. Relation between human and non-human world on a mutually arguing platform of respect and love.

5. The actions or agenda of actions for sustaining the "nature".

Gandhian thought understands the above listed points and importance of holism and common good. Gandhi's concept of truth and non-violence is equally applicable to environmental issues. The intermediate simple living and high thinking of man with oneness of all creation on platform of truth and non-violence associated with anarchistic spirit underlines the Green political thought in Gandhism.

For maintaining the nature and sustaining it for cause of survival principles of green political structure advocates decentralization. "The greens seek to substitute for all that a decentralized, democratic, society with .... Political economic and social institutions locating power on the smallest scale (closest to have) that is efficient and practical. There guiding principle is that no authority be held at higher level than is absolutely necessary[31]. The decentralization, by virtue of its position avoids violence. Weebs nature and promotes self suffice community and civil sustainability. Gandhi is of firm opinion, quite a advance to appearance of Green political action, when he recognizes the anarchistic idea of state with locally sustainable villages Gandhi opines" That state is perfect are governed where the people and vogerned the least. The nearest approach to purest anarchy world be a democracy based on non violence" [32] He, however is anxious about this state, so opines, "I do not conceive of such a golden age. But I do believe in the possibility of a predominantly non-violent society. [33]

Gandhi says I look upon an increase in the power of the state with the greatest fear, because although while apparently doing good by minimizing exploitation, it does the greatest harm to mankind" The eccentricity of power causes exploitation and so Gandhi believes in grass root level participatory evolving democracy. "True democracy cannon be worked by twenty men sitting at the center. It has to be worked from below by the people of every village" [35] By a participatory democracy the people will understand their own problems very clearly and because they are ethical person with "essential economy" only the ecological concern get secured automatically.

Gandhi believes not only in decentralised political power but also in decentraliased economy with giving dignity to labour. "You cannot build non-violence on a factory civilization, but it can be built on self contained villages ... Rural economy as I have conceived it eschews together, and exploitation in the essence of violence. [36]

Thus it is seen that the Green political theory, which is ethical in its nature assumes the human as part of nature for cause of survivalism and justice to coming generation, is saying nothing new apart from Gandhian ideas.

Gandhi never said explicitly on political or economic philosophy nor elaborated the Green political issues or environmentalism, but his ideas were to advanced in their virtue that the Green concern find their abode in his thought. The Green political theory which tries to balance human claims and appeals with non-human world gets settle when Gandhian ethical man acts on truth and non-violence for cause of common good under patronage of local civil society with ethical economics.

In the passing references it is also important to note that Gandhian Green political theory has advocated not only a self-disciplined human and village but also the welfare of disabled and animals. Gandhi's "Sarvodaya stands for the "Welfare of all". The welfare of human world and animal world are very important. [37] Gandhi's concern of non-violence pulled everyone near to each one and put under the same umbrella. Gandhi's theory of life does not compartmentalise the existence into any faction, class or groups but it ushers the whole beings and non beings in one pot. This is the unique feature of Gandhian philosophy as well. Gandhi's faith in non-violence[38] is the faith ubiquitous consciousness of almightily, and God is most natural and "eco" in nature therefore.

REFERENCES

1. MK Gandhi, Harijan, 2/3/40, p. 23

2. MK Gandhi , Young India 31/12/31, p. 427-428.

3. MK Gandhi, Harijan 14/3/36, p. 39.

4. Shreeman Narayan, The selected works of Mahatma Gandhi, Vol VI, Navjeevan Publishing House, Ahmadebad, 1968, p. 154.

5. Lester Brown, A planet under stress in John S Dryzeck and David Schlosberg, Debating The earth, Oxford University Press, 2005, p. 37.

6. MK Gandhi Young India, 24/6/26, p. 230.

7. Ibid p. 230.

8. MK Gandhi, harijan 5/9/36, p. 40.

9. Andrew Vincent, Green Political Theory in Richard Ballamy and Anderw Mason, (ed) Political concepts, Manchester University Press, 2003, p. 182.

10. MK Gandhi young India 13/11/24 p. 378

11. MK Gandhi Young India 26/9/29 p. 320

12. MK Gandhi young India 13/11/24 p. 378.

13. Andrew Vincent, op. cit p. 184.

14. Andrew Dobson, Environmentalism and Ecologism in Alan Finlayson (ed), contemporary Political thought, Edinburgh University Press, 2003, pp. 364-365.

15. MK Gandhi, Harijan, 29/8/6, p. 226.

16. Shriman Narayan op. cit p. 328.

17. MK Gandhi, Harijan 1/3/46, p. 285.

18. Pyavel al, towards, New Horizon, Navjeevan Publishing House, Ahmadabad, 1959, p. 86-87.

19. MK Gandhi, Harijan, 23/2/47, p. 39.

20. Andrew Heywood, Political Ideologies, Palegrave Macmillan, 2003, pp. 268-269.

21. Lester Brown op. cit pp. 37-40.

22. MK Gandhi Young India, 12/11/31, p. 355.

23. MK Gandhi Young India, 15/4/26, p. 142.

24. MK Gandhi Young India 17/6/26, p. 218.

25. World's commission on environment and development, our common future Oxford University Press, 1987, p. 43.

26. James meadowcraft, Sustainable development a new (ish) idea for a new cenntry, Oxford University Press, 2005, p. 270.

27. MK Gandhi, Harijan 2/2/47.

28. The diary of Mahadev Desai Vol I, p. 221, 10-7-32 in Shriman Narayan Vol. V Navjeevan, Ahmedabad, 1968.

29. Collected works of Mahatma Gandhi, Vol X, 2/4/10, p-206, in Shriman Nayaran, ibid, p. 431.

30. Andrew Dobson, Environmentalism and ecologism Ibid, p. 377.

31. Robert Goodin, Principles of Green Politics in Alan Finlaysun, Ibid p. 399.

32. MK Gandhi, Harijan 21/3/40, p. 211.

33. MK Gandhi, Harijan 9/3/1940, p. 31.

34. NK Bose (ed) Selections from Gandhi Navjeevan Publishing House, Ahmedabad, 1957, p. 41.

35. MK Gandhi, Harijan, 18/1/48, p. 519.

36. MK Gandhi, Harijan, 4/11/39, p. 331.

37. Savita Singh Bharti Mishra (ed) Gandhian Alternative Vol II, Concept Publishing Company New Delhi, 2005, p. 191.

38. Ibid pp. 84-91.

CHAPTER-19

Gandhian Consciousness of Non-Violence in Green Concern

Gandhi is not a man. It is an idea a of non violence thought, a concept and a philosophy. Gandhian soul is not a something mystic supra human entity; it is infact the basic edifice of socio-politic economic and ecological structure of society. Gandhi never claimed a political philosopher in himself but Gandhi claimed that "I have applied it in every walk of life, domestic, institutional economic and political. I know of no single case in which it has failed. [1] Gandhi is proponent of action not of theory: By being a man of action, he reacted with vigour to every critical situation of political social, religious and cultural conflict during his time and tried successfully to resolve it by truthful and non-violent means.[2]

The structure of society rests with same assumption. The fundamental assumption is live and let live. The theories of state formation and going ahead from this milestone, the theory of human species notifies the mutual solidarity cohesiveness be it any state – of community from city state to civil societies, the force of accepting other have been working. The nature of man has been portrayed by different philosophers in different manner but for Gandhi "Man's nature then is not "Himsa", but "Ahimsa", for he can speak from experience his innermost conviction that he is not the body but Atman"[3]

The basic unit of society, that gives meaning and ontological expression to milieu around him is not a "body" but a "soul". The very soul which is non-violent even Thomas Hobbes dictation about human, nature which indicates 'War of every man against every man" or life of man is solitary, poor, nasty, brutish and short[4] finally ends in peace and cooperation for the cause of self preservation. Thomas Hobbes opines" Therefore the law of nature is the dictate of right reason, conversant about those things which are either to be done or ormitted for the constant preservation of life and members, as much as in us lies.[5]

The final outcome and the fundamental value is, thus a mutual existence and avoiding those forces which are detrimental to life. Now if we shift our attention to the basic postulate of Green political theory it is the very survival of the whole natural ecosystem of the planet" [6]

The devasted ecology and unbalanced environment have put the big question of survival of humanity and other creations. The value of survivalism is, now of top ranking concern so in consequence. "The Green political theory addresses the condition for our biological continuance as a species." [7]

The basic question here is that any act which is detrimental to survival is violence and thus is not allowed in green political ideas.

Gandhi's non-violence is the love of the human race. It is infinitely greater than and superior's brute force[8]"

The concerns of ecocentric school is ruthless exploitation, violent milching and brutal extration of natural resources, flora and fauna by human being for the cause of satisfying then. Unlimited ever growing needs and desires and ecocentrists want to stop it immediately just for cause of preservation, and continuance of biotic and abiotic existence. Gandhiji seems standing behind this grave concern when he echoes" Scientists tell us that without the presence of the cohesive force amongst the atom that compromise this globe of ours, it would crumble to pieces and we would close to exist. And even as there is cohesive force in blind matter so must there be in all things animate, and the name for that cohesive force among animate being is love.[9]

Gandhiji's concern of love to all is not only limited to live beings but also to whole creation. Gandhi continues "When non-violence is accepted as the law of life; it must pervade the whole beings and not be applied to isolated acts."[10] Gandhi seems much advanced to those western political thinkers or german green party who later on underlined the law of life of all being be it green or whole non-green world. Because Gandhi sees the truth as God and finds "Truth is the sovereign principle, which includes numerous other principles"[11] then he continues "Truth is God – nothing else, nothing less[12]". He keeps on saying "the nearest approach to truth was through love"[13]. 'Truth is God', gives me the greatest satisfaction and when you want to find the truth as God the only inevitable mean is love, i.e.; non-violence" Thus Gandhi accepts love as the presence of truth as God. At every place. He opines "Ahinsa means universal love."[14]

The Green political ideas have given a major rule to mother nature. They have assumed natural planet as a single bigger organism and human being a integrated sub-organ to the same. Because green political thinkers put the whole creation in one entity, the mutual respect and harmony and non violence are pre condition of the oneness. This oneness ultimately drives towards non-violent attitude necessity from human beings. Gandhi says "Ahimsa includes the whole creation, and not only human"[15].

It is not out of the subject of the topic if we understand the major current of Green political idea. This idea has been projected from two sides the first in ecocentrism and other is anthropocentrism. The central philosophical axiom of this perspective is that there is no firm ontological divide in the field of existence[16]. From ecocentric point the planet is a single organism and the whole creation has an intrinsic value. The men is simply a moral agent which has to act for cause of whole organism because the other integral part, i.e. non-human world has its own appeal and say. Nature itself is thus portrayed as an ethical community, meaning that, human beings are nothing more than plain citizens who have no more rights and are no more deserving of respect than any other member of the community. Such a moral stance implies bio-centric equality, the principle that all organisms and entities in the ecosphere are of equal moral worth, each of them beings part of an interrelated whole[17]. Arne Naess[18] expressed this consciousness as an 'equal right to live and bloom'. Though food chain and biological survivalism and social Darwinism limit this subjection on practical grounds, even than if we see behind this philosophy, it is nothing but the axiom of non-violence and Gandhi quite aptly advances his view "It is not through strife and violence, but through non-violence that man can fulfill his density and his duty to his fellow creatures. At the center of non-violence is a force which is self aching.[19] Gandhi cares for all the creation. The practical solution to this stage where some non-violence is unavoidable, Gandhi opines for self-suffering civil communities. Gandhian anarchism points towards the stateless civil society. The community members of such society underand their responsibilities and duties. The rights they claim is only service to others. Gandhi's view was willing submission to social restraint for the sake of well being of the whole society enriches both the individuals and the society of which one is a member. The only individualism that is justifiable that which is built up in the service of others,[20].

So we find that deep ecocentrism may point the perfect absence of non-violence but beings adjusted with practical condition of survival, this stage is not possible. The notion and understanding of least of the least non-violence finds a solution in theory of trusteeship. Gandhi understands that perfect non-violence is not possible because for survival things reasources and means will have to be used, but over use and exploitation will lead to marked-non-violence. Gandhi then puts the idea of trusteeship. "I am imiting those people who consider themselves as owner today to act as trustees, i.e., owners not in them own right, but owners, in the right of those whom they have exploited" [21] But Gandhi understands that theory of trusteeship is not reached perfectly so he is agree to a minimum ownership of state[22]. As the state represents violence so far practical purpose the violence is allowable.

Gandhi's idea abridges the gap between ecocentrism and anthropocentrism. The other dimension of Green political theory as the name suggests that anthropocentrism. Stresses :that human beings are the sole criterior of value" [23] Green political theory itself understands that deep anthropocentric values are outside of the kernel of thought because of its utter eccentricity on human and on this ground impracticality of the same, pliant authropocentrism stresses co-dependency with nature, although still filtered through human interests. The pliant perspective leaves traditional normative theory largely unchanged. However new issues and questions are mapped on to the other normative concepts. Green political theory thus takes conventional issues of justice, equality citizenship or rights and then adds a green dimension, emphasing co-dependency with nature" [24]. In light of normative concepts thus Green political theory balances the human with "nature" and in this way allows that minimal violence that is essential for survival.

This is the very platform which Gandhi sees in minimal state steewardship. The notion of violence in green political theory gets manifestation not only in exploitation of present resources flora, fauna and non human world but also in the existence and claims of present generation and forthcoming generation. The traditional concepts of political philosophy, i.e. justice, state, rights, liberty, equality etc. have the element of subtle violence in them as enshrined value but this violence has been made non-violant by colouring the existence and appeal of ecology as a whole.

This is the same platform which Gandhi wants to realize through his concept of non-violence – the other face of truth coin. Gandhi wants a self sufficient, non-violent civil society. 'To Gandhi, Civil society means a new social order, which is a based on mutual respect and tolerance, open and secular institution where people will learn the lesson of accommodating together, an arena of self-rule or self-management based on self-sacrifice for the pursuit of common good. All these will be based on the truth and non-violence on the hand and the notion of cooperation and social feeling on the other."[25]

Going in direction of ecologism and green sphere the mutual balance must be not only in human world but also between biotic and abiotic world because it starts not from a conception 'humanity' or human needs, but from a vision of nature fragile relationship between living species-including the human species-and the natural environment. Human kind no longer occupies center stage, but is regarded as an inseparable part of nature." [26]

Gandhi's "non-violence is soul force on the power of the God-head within us. We become godlike to the extent we realize non-violence[27] and non-violence is the law of our species, as violence is the law of brute" [28]

A brute is never welcome in Green political ideas, even the present production and environmentalism which adjusts the human needs and desire with non-human world appeal and claims rests on management[29] not on any coercion or brutal thrust. At the same time ecologism which stresses radical change in our relationship with the non-human world and in our mode of social and political life, unanimously rests that "radical green politics is itself a spiritual experience in that it is founded on a recognition of the 'oneness' of creation and a subsequent 'reverence for one's own life, the life of others and the earth itself." [30]

It is quite pleasing to note that Gandhi's soul has witnessed this all in name of non-violence much earlier than the environmentalistic or ecologistic movement. "Truth and non-violence are the two cardinal principles of Gandhism. Non-Violence sounds negative non-killing, non-hurting, but Mahatma's non-violence is a positive concept standing for love and goodwill to all[3]. Gandhian spirit of non-violence is the cementing force of economics with environment material development with moral development and machine with mankind. The ecocentricity of Gandhi underlines the non-violence in any object or subject. The true non-violence to non-human world will lead to true political philosophy of ecocentrism.

REFERENCES

1. MK Gandhi, Harijan, 6.7.1940, pp. 185-186.

2. Sumit Kr pathak, Gandhi and Civil Society. The Indian journal of Political Science, Vol LXIX, No. 2 Apr June 2008, p. 269.

3. MK Gandhi Young India 24.6.1926, p. 230.

4. George H. Sabine, A history of Political thought, George G Hararap and Co. Ltd. London, 1941, p. 464.

5. Thomas Hobbes Deceive ch 2, 1. English works, Mokesworth (ed) vol. II p. 16, quoted by Gitt Sabine op cit, p. 465.

6. Andrew Vincent, Green Political Theory in Richard Bellany and Andrew Mason (ed), Political concepts Manchester University Press, Manchester, 2003, p. 182.

7. Ibid p. 182

8. MK Gandhi, Harijan 5/9/1936, p. 236.

9. MK Gandhi Young India 5/5/20, p. 7.

10. MK Gandhi Harijan 5/9/1936, p. 236.

11. Ibid p. 236

12. Shriman Narayan, The selected works of Mahatma Gandhi, Vol VI, Navjeevan Publishing House, Ahemdabad, 1960, p. 97.

13. MK Gandhi young India, 31.12.31, p. 427-428.

14. Shriman Narayan op. cit p. 153.

15. MK Gandhi Harijan 14/3/36, p. 39.

16. Warwick Fox quoted in R sylvan, A critique deep ecology, radical philosophy 40 & 1, 1984-85, p. 103.

17. Andrew Heywood, Political Ideologies, Palgrave Macmillan, 2003, p. 281.

18. Arne Naess, Community and lifestyle : outline of an ecosophy, Cambridge University Press, 1989.

19. MK Gandhi, Harijan, 14/3/36, p. 39.

20. Biswas SC (ed) Gandhi Theory and practice social impact and contemporary relivance Indian Institute of advanced study Shimla, in association with KP Baghi & Company Calcutta, 1969, p. 322.

21. MK Gandhi young India, 26/11/37, p. 368-69.

22. Shriman Narayan, Ibid, p. 370.

23. Andrew Vincent, Ibid p. 183.

24. Op. cit p. 184.

25. sumeet K. Pathak, Ibid p. 276.

26. Andrew Heywood, Ibid pp. 268-269.

27. MK Gandhi Harijan 12.11.36 p. 326.

28. MK Gandhi young India 11.8.20., p. 3.

29. Andrew Dobson, Environmentalism and ecologism (2000), in Alan Finlayson (ed) contemporary political though, Edinburgh University Press, 2003, p. 364.

30. Andrew Dobson ibid p. 377

CHAPTER-20

Gandhian Concern in Ecoanarchism

Green political theory is a new arrival in realm of political theory. Though the accounts of environmental homage and ecological concerns are found in writings of greek political theory and ancient Indian Vedic political thought yet a systematic study of this stream of intellect emerged in post mid period of twentieth century. Hyper industrialisation, merchanisation, urbanization victory of "man" on "innocent" nature ruthless destruction of national flora and funa and violent exploitation of natural resources awoke the high concerns for maintaining the nature just for the cause of survival of human species. "Green political theory claims to be focused on the very survival of the whole natural ecosystem of the planet. In consequence, it also addresses the conditions for our biological continuance as a species".[1] The unique signature of this green political theory its subtle relation with the established concepts of state, morality, democratic values, rights freedom, equality, justice etc. "The green political theory conceives of itself as "green", "environmental" or "ecological" because of its key focus on nature.[2]". The nature is seen as a crucial entity in its own right – of which we are a just minor part. Thus the green political theory articulates the ecology and biological survival by keeping the human species as an integral part of nature.

The Green political theory emphasises two extremes. On one hand the ecocentrism conceives the ecology in center; the eco-centric value perspective has developed on two lines – The first is intrinsic value theory which sees nature as an end in itself and the second is the psychological and ontological view of human species to it on the other anthropocentric value perspective puts the human beings as the sole criterion, however the pliant anthropocentrism stresses co-dependency with nature.[3] Thus green political theory takes in cognizance the nature and then adds up the political concepts with it just for survival of better citizenship. In the same light the ideology of ecology represents a new style of politics. It starts not from a conception of "humanity" or human needs, but from a vision of nature as a network of precious but fragile relationship between living species – including the human all these concerns in name of non-violence existed much earlier than the environmentalistic or eco-logistic movement.

"Truth and non-violence are the two cardinal principles of Gandhism. Non Violence sounds negative : non-killing, non-hurting. But the Mahatma's non-violence is a positive concept standing for love and good will to all[3]. Gandhian spirit of non-violence is the cementing force of economics with environment, material development and machine with mankind. The eco-centricity of Gandhi underlines the non-violence in any object or subject. The true non-violence to non-human world will lead to true political philosophy of ecocentrism. Species and the natural environment human beings are therefore required to practise humility, moderation and gentleness and to abandon the misguided dream that science and technology can solve all their problems."[5] Thus it is clear that the ecology in the central thought of green political theory and it takes into account the nature – the wild nature-which in corporates the "human species" as an integral part of the same. It respects the nature. It also respects the human species and its actions, needs etc. but at the same time the rights and coexistence non-human live or un live creations are there.

In this context the word eco-anarchism is a new development in Green Political Theory. Bookchin[6] has opened that there is a functional relationship between the ideas of anarchism and ecology. Anarchists believe in stateless society where mutual relationship and solidarity come in play. The ecology talks about the mutual relationship between abiotic and biotic world, where the "natural" bio-centric justice will prevail.

Econarchism finds its root in writing of Prince Kropotkin. Kropotkin asserts that anarchism is a principle of life and conduct. Under which society is conceived without a govt., harmony in such a society beings obtained not by submission to law or by obedience to any authority, but by free agreements conducted between various group, territorial, and professional, freely constituted for the sake of production and consumption as also for the satisfaction of the infinite variety of needs and aspirations of civilized living[7]. Tolstoy, Warsen, Tucker, Schimidt, Sturner are some other champions of anarchism but the anarchy communism is the principal light giver of eco-anarchism.

Anarchists oppose the state and its authority, in the same way that is unauthority of man ecologism believe in natural rule not the rules or ruling with force human on nature. Human rule is exploiting and violent, moreover it is anthropocentric which, again is opposite to natural wholism. On the vital ground of unnaturalness, morality, unjustification of ruling with power instead of mutual cooperative will shakle the principles of diversity, balance and harmony. Bookchim "linked an anarchist community to the eco system. Kropotokin attempted to provide a biological foundation for social solidarity. He argued the unsuitability of Charles Darwin's theory of evolution as mutual conflict can never be the base of evolution; rather collecting collective energies through cooperation mutual aid, respect and harmony strengthens the evolutionary process. He sees the mutual aid as the principal means of human and development[8].

Anarcho-communism which enlightens the eco-anarchism admires self sufficient. Communities. These communities look on the principle of sharing and collective endeavor. They are bound with compassion and solidarity. Self has society s their endeavour. Participatory democracy along with face to face interaction is the way of action in such anarcho communistic societies[9]. The ecologism believes in basic themes of ecology, solism, sustainability environmental ethics, self actualization[10]

If we analyse the ecological themes we will find that the basic assumptions of anarcho-communism are there in vast coverage of ecologism. The self sufficient decentralized communities, encourage balance between human beings and natural world. [11]

It is also clear that in eco anarchistic society the human needs are also subject to the naturalistic harmony with wild nature and other living being, so exploitation of environment for materialistic show of will shed away and a true spirituo-natural order of ecological survivalism will appear.

The political ecologists stresses two points : one, that consumption of material goods by individuals in advanced industrial countries should be reduced; and two linked to first, that human needs are not best satisfied by continual economic growth as we understand it today[12]. Thus reducing consumption is the important theme of ecologism-which is also part and parcel element of eco-anarchistic thought.

Thus ecoanarchism has following central themes.

1. Stateless society or minimum statism.
2. Mutual harmony between human and non-human world.
3. Self Sufficiency.
4. Decentralisation.
5. Participatory decision-making.
6. Spiritual outlook of life.
7. Ethical life, and finally.
8. A deep respect to ecology and nature where human is also integral part of the same.

Now we can analyse Gandhian Philosophy and his agenda of action in fabrics of eco anarchism. First of all it will be interesting to know that in Indian political thought after Manu Gandhi stands as an staunch advocate of anarchistic society. He does not believe in the oppressive power of state as it is a form of organized violence. Gandhi writes "The state represents violence in a concentrated and organised form. The indivudual has a soul but as the state is a soulless machine, it can never be weaved from violence to which it owes its very existence"[13] It is to be noted down that ecoanarchistics do not believe in any type of violence. Gandhism is, thus true replica where any type of violence be it any form is strictly outcasted. Gandhi opines that non-violence is name of absense of any malafides to all beings. Thus non violence in its executive way is bonafide good wishes to all beings. This is pure love[14]. Gandhism can never support any violence as it is pure love so this extends to all the creation the nature can, thus, never be inflicted with the violent stakes. The "beings" as referred above are natural integers Gandhi further elaborates that "Truth is God" gives me the greatest satisfaction. And when you want to find truth as God the only inevitable means is love i.e. non-violence and since I believe that ultimately the means and end are convertible terms. I should not hesitate to say that God is love. [15]"

Gandhi's non violence is thus love in another aspect and because truth is eternal so non violence thus in the above context, gets linked with almighty. For eco anarchists the element of non-violence is value addition. Green thought does not want ruthless violent behaviour, as for them environmental justice affirms the sacredness of mother earth, ecological unity and the interdependence of all species, and right to be free from ecological destruction[16]. Alike the deep revence to ecological in heart of ecoanarchists Gandhi respect every life in shape of God. "My doctrine means, I must identify myself with life, with everything that lives, that I must share the majesty of life in the presence of God. The sum total of this life is God" [17] Gandhi's self consciousness of "manhood" is spread in all forms of life so a Gandhian can never be brutal. Gandhian man loves everyone "where there is love, there is life, hatred leads to destruction" [18] Gandhi's love is the cohesive force. He goes on saying "Scientists tell us that without the presence of the cohesive force.

Amongst the atoms that comprise this globe of ours, it would crumble to piece and we would cease to exist. And even as there is cohesive force in blind matter so must ther be all thigns animate: and the name for that cohesive force among animate bring is love" [19] Gandhi thus spreads mutual respect, love and harmony. He is staunch advocate of harmonious existence of manhood. Though same time Gandhi appears to be an anthropocentric when he echoes that "The supreme consideration is man" [20] yet this echo should be analysed from the point of view "Man is not a brute. He has risen to a higher state after countless births in brute creation.... Bestiality is as far removed from manhood, as matter from spirit.[21]" It is precipitated that Gandhi advocates for a wonderful shift from anthropocentric school to ecocentric sphere on the trains of spirituality and harmony. Gandhi does extra ordinary work when he combines "man" and wild nature life on the platform of infinite love and mutual respect.

Alike ecoanarchists Gandhi believes in interdependent but self-sufficient life. Gandhi's Economic activities are not for material outlook or conspicuous consumption. He is not favouring those economic or political efforts which can not do good for others. :That economics is untrue which ignores or disregards moral values[22]. True economics, on the other hand, stands for social justice, it promotes the good of all equally including the weakest and is indispensable for decent life[23]. It will be agreed upon unanimously that the ecological problems and environmental degradation have come out of the materialistic lust and desires – which certainly are not morally sound if abiotic world is correlated with the anthropocentric angle.

Gandhism belief in plain living and high thinking leads the man in ecoanarchistic state – the state where self sufficiency is there. "The ideal of creating and unlimited number of wants and satisfying them seems to be a delusion and snare[24]" The divine and moral nature of the man, as anarchists understand, may flow in Gandhian way of thinking.

The decentralized economy and participatory decision making have taken a wonderful culmination in Gandhian thought. Gandhi says, "You can not build non violence on a factory civilization, but it can be built on self contained villages .... Rural economy as I have conceived it, eschews exploitation altogether, and exploitation in the essence of violence[25]. Gandhi goes on saying further "My purpose is to present an outline of village govt. Here there is perfect democracy based open individual freedom. The individual is the architect of his own govt. The law of non-violence rules him and his government he keeps on maintaining that "A society organised and run on the basics of complete non violence would be the purest anarchy ...., Yes. It is realizable to the extent non-violence is realizable. That state is perfect and non violent where the people are governed the least.[27] However Gandhi is doubtful about such a govt. which will be entirely non violent he says "A govt. cannot succeed in becoming entirely non violent, because it represents all the people" [28], In such condition " the nearest approach to purest anarchy would be a democracy based on non-violence" [29]. Gandhi thus believes the self regulating, independent villages where the moral spiritual and non violent man will be the ultimate unit and this unit will be harmonious to every affair of life. Gandhi asserts that independence must begin at home. Thus, every village will be a power. It follows, therefore, that every village has to be self-sustained and capable of managing its affairs. Such a society in necessarily highly cultured in which every man and woman knows what he or she wants[30].

Gandhi was aware of exploitation and its relation with industrialization. Ecological movements and eco anarchism have identified that hyper industrialisaction has depleted the environment and thus destroyed the ecological balance. Gandhi says "Industrialism, is, I am afraid, going to be a curse for mankind.... Industrialism depends entirely on your capacity to exploit"[31]. It at the same time, is there, that Gandhi was against only to those machines which are obstacles in human labour practices but "that machinery is lawful which subserves to interest of all[32] "I am not against machinery as such, but I am totally opposed to it when it masters us. [33]"

Thus it is seen from above accounts that Gandhi's idea of society resembles with the paradigm of ecoanarchism and its assumption. Gandhi's philosophy is a wholistic philosophy which underlines the divine, moral and spiritual nature of and the very man is to live in complete harmony and mutual respect by taking only that much from society which is due and necessary to him. Gandhi's man bases on the theory of trusteeship. Gandhi's idea is that "Everything belonged to God and was from God.... When an individual had more than his proportionate portion he become trustees of that portion for God's people[34]". Gandhi's industrious person may earn according his abilities but after the need full as per bare necessities, the remaining will be the share for others, for goodness of others and society. This whole industrious effort will be based on non possession attitude of the man[35] Gandhi's theory of trusteeship hints out specifically the line of action that is based on attruistic economics. As has been discussed earlier that Gandhi has bridged the gap between anthropocentric and ecocentric man by assuming the "man" as divine, ethical attruistic unit that is working for goodness of others.

Thus we see that a complete concern of eco anarchism finds its place in Gandhian thought. Infact the superiority of Gandhiam thought lies in the fact that the ecological concerns and the welfare of green political theory with special reference to eco anarchism find their pivot place in self sufficient, spiritual, humble. Non-exploiting, non violent, truth worshiping man the very unit of nature which is integral to nature, and Gandhi. Finally, sees the God (truth is God) in whole nature by echoering "I find in the practice of tree worship a thing instinct with a deep pathos and poetic beauty. It symbolises true reverence for the entire vegetable Kingdom, which with its endless panorama of beautiful shapes and forms, declares to us as it were with a million tongues the greatness and glory of God. [36]

REFERENCES

1. Andrew Vincent, Green Political Theory, in Richard Bellamy and Andrew Mason (ed) Political concepts, Manchester University Press Manchester and New York, 2003, p. 182.

2. Ibid p. 182.

3. Ibid pp. 183-184.

4. Smriti Singh, Bharti Misra, (ed) Gandhian Alternative vol II, concept Publication, New Delhi, 2005, p. 122.

5. Andrew Heywood, Political Ideologies, Palgrave Macmillan, New York, 2003, pp. 268-269.

6. M Bookchin, Anarchism and ecology in 9 woodcock (ed), The Anarchist Reader Fontana, London, 1977.

7. Prince Kropotkin, Mutual aid, Boston, Mass Porter sergeant, 1914.

8. Andrew Heywood, op. cit pp. 202-203.

9. Ibid p. 203.

10. Ibid p. 269.

11. Ibid p. 288.

12. Andrew Dobson, Environmentalism and ecologrim, finlaysin (ed) contemporary political though, Edinburgh University Press, Edinburgh, 2003, p. 375.

13. Young India, July 2, 1913, p. 162.

14. Young India, March 9, 1920.

15. Young India, 31/12/1931, pp. 427-428.

16. Principles of Environmental justice, adopted at the first national people of colour environmental leadership summit 7/10/91 Washington D.C.

17. Harijan, 15/2/1948, p. 33.

18. Young India 5/5/1920, p. 7.

19. Ibid p. 7.

20. Young India 13/11/1924, p. 378.

21. Young India 29/4/1926, p. 157.

22. Young India 26/12/1924, p. 421.

23. Harijan, 9/10/1937, p. 292.

24. Harijan, 29/8/1936, p. 226.

25. Harijan, 04/11/1939, p. 331.

26. Harijan, 26/7/42, p. 238.

27. Harijan, 21/7/40, p. 211.

28. Harijan, 9/3/40, p. 31.

29. Harijan, 21/7/40, op. cit.

30. Harijan, 28.7.1946, p. 236.

31. Young India 12.11.1931, p. 355.

32. Young India 15.4.1926. p. 142.

33. Harijan 27.2.1937, p. 18.

34. Harijan 23.2.1947, p. 39.

35. Jawahar lal Dwivedi, India Political thought (Hindi) Hindi Medium Execution Directorate, Delhi University, 1998 pp. 141.

36. Young India, 26/9/29. p. 320.

BIBLIOGRAPHY

1. Ageyman, Julian, Bullarld Robert and Evans Bob (2003) eds, just sustainabilities, Cambridge, MA MIT Press.

2. Alln J. Braham P and Lewis P (1992). Political and economic forms of Modernity. Cambridge, Polity Press

3. Anderw Kenneth J et. al. 1995, Economic Growth carrying capacity and the environment, science 268, (28 april) : 510-1

4. Beekerman, Wilfred (1974), In defence of economic growth (London, cape)

5. Backerman, Wilfred, A poverty of reason sustainable development and economic growth, (Oakland, CA: The Independent Institute)

6. Bookchin Murray (1980), towards an ecological society (Montreal ; Bleck Rose)

7. Bookchin Murray (1982) The modern crisis (Philadelphia PA: New Society)

8. Bookchin Murray (1988). Social Ecology v/s Deep Ecology, Socialist Review 18(3): 9-29

9. Bookchin Murray (1990), The philosophy of social ecology (Montreal : Black Rose)

10. Bullard Robert D (1994) ed confronting environmental Racism : voices from the Grassroots (Boston : South End Press)

11. Bryant Bunyan (1995) ed Environmental Justice : Zosus Policies and Solutions (Covelo, CA: Island Press)

12. Camacho David (1998) ed, Environmental Injustice : Political struggle (Durham NC, Duke University Press)

13. Catton William R (1980) overshoot : The ecological Rasis of Revolitionary change (Urban I2, University of Illinois Press)

14. Clark John (1990) Renewing the Earth; The Promise of social ecology: A celebration of work of Murray Bookelin (London, Green Print)

15. Dobson Andrew (1988) justice and the environment : Conceptions of Environmental sustainability and social justice, Oxford University Press.

16. Dry Jeck John, Downes, David, Hunold Christian and Schlosberg David (2003), Green States and Social Movements Oxford University Press.

17. Eastern Brook Gregg (1995) A moment on Earth: The coming Age of environmental optimism (New York: Penguin)

18. Eckersley, robyn (1985) (ed) Markets, the state and the environment : towards integration (Melbourne - Macmillan)

19. Ferry Luke (1995) The New Ecological order (Chicago ; University of Chicago Press)

20. Fox, Warwick (1990), Towards a transpersonal ecology : Developing New Foundation for Environmentalism (Boston MA : Shambhala)

21. Gorz Andre (1980), Ecology as politics (Boston South End)

22. Harvey David (1996) Justice, Nature and Goegraphy of difference (Oxford : Balckwell)

23. Jacobs Michael (1981) The green economy environment sustainable development and the politics of the future (London : Pluto)

24. Mason, Michael (2000), Environmental Democracy (London ; Earthscan)

25. Mathews, Freya (1996) ed ecology and democracy (Portland or : Frank Cass)

26. Meis, Maria and Shiva Vandana (1993), Ecofeminism (London : zed)

27. Plumwood Val (19930, Feminism and the Mastery of nature (London ; Rootledge)

28. Redclift, Michael (1987) Sustainable Development : Exploiring the contractions (London : Methuen)

29. Schnaiberg, Allan (1980), The environment ; from surplus to scarcity oxford (Oxford University Press)

30. Shiva Vandana (1995) Stolen Harvest : The Hijacking of the Global food supply (Boston : South End Press)

31. Simon, Julian (1995) ed, The state of Humanity (Oxford: Blackwell)

32. Tokar Brian (1987), The Green Alternative (San Pedro, CA: R & E Miles)

33. VK Natraj Kamlesh Mishra Neeru Kapoor (2005) Gandhian Alternative IV volumn Concept Publishing, New Delhi.

34. Winssenburg, marcel (1998), Green Liberalism : The free end and the Green Society (London : UC & Press)

35. Young's (2000) The emergence of ecological modernization : Intgrating the Environment and the economy (London Routledge)

◆◆◆